A MANAGER'S GUIDE TO TELECOMMUNICATIONS

Martin Gandoff

Heinemann: London

William Heinemann Ltd
10 Upper Grosvenor Street, London W1X 9PA

LONDON   MELBOURNE
JOHANNESBURG   AUCKLAND

First published 1987
© Martin Gandoff 1987

British Library Cataloguing in Publication Data
Gandoff, Martin
   The manager's guide to telecommunications.
   1.Office practice – Great Britain – Automation
   I. Title
   651   HF5548.2

ISBN 0 434 91068 6

Typeset by Express Typesetters Limited, Farnham
Printed by R. J. Acford, Chichester, West Sussex

Contents

Contents

List of illustrations

Chapter 7 *Local Area Networks (LANs)*

Appendix 1 *Mercury*

Appendix 5 *The ISO 7-layer model*

Preface

While researching material for this book, I realised very quickly that I had set myself an extraordinarily difficult task. It is never easy to do a book covering a dynamic area, but with telecommunications (transfer of information over a distance) both for voice and data, the last couple of years have seen the technology expand in leaps and bounds and with it, the applications or modifications to existing applications.

In my view, any attempt to produce a book that is not out of date before it is read can only be successful if it sets out to educate managers in the basic principles behind the various techniques and encourages them to look more deeply into what is available.

In accord with well defined management practice, good managers need enough background in a subject to be able to brief subordinates so that they know what to do or what to look for, and then to be able to appreciate what they return with so that the right decisions can be made. So, apologies for the high 'teaching' content.

I have been very careful not to distinguish between telecommunications (telecomms), which traditionally has meant voice communication, and data communications (datacomms). As separate areas, they often come under the control of a telecommunications manager on the one hand and a data processing manager on the other. I believe that the distinction is now really rather pointless, since, as will be discussed, the two types of communication often use the same technology, the same networks and often even the same wire. The buzzword for this meeting of voice and data is 'convergence'.

In order to keep the book to a manageable size I have had to omit several areas; this will probably annoy some readers. In particular, there is no coverage of answering machines, paging equipment and video.

For the subjects of Chapters 4–6, I have discussed the relevant theory, broadly reviewed the range of facilities and then given examples of hardware and software. You will note that I have made almost no recommendations as to specific products – I never intended to produce a buyer's guide.

I hope that by the time you have read the book, you will be able to answer at least partially, questions such as:

- This sounds interesting, what is it?
- What can I do to find out more about it?
- What can it do?
- How does it work?
- How will it help me?
- How do I find out which one to select?

At the back, you will find a fairly long glossary of terms. I make no apology for this. There are so many buzzwords and terms in use, many of which are similar, that a quick-reference dictionary like this one would have helped me considerably.

I would like to express profound gratitude to everyone that gave me help in the form of product information, customer education material, photographs and a lot of their time: in particular, Northern, Mercury, Easydata, Effem and John Smart of Manners, Borkett and Partners (who handle some of BT's PR). Especial mention for Gloria for so much moral support during the

11

What does the manager want to know? What does he want to be told? Can I avoid teaching him? No I can't – he needs to know much more and, hopefully, I know more than him.

The subject is incredibly wide and certain concepts/techniques will bubble to the surface almost immediately.

BT Bless them! They can't or won't tell you all about their services in a structured way. It seems that there is a very wide range of services based initially on different networks for different applications. We must tell people about the main functions/workings of PSTN and PSS. What is the set-up behind the telephone handset – PABX, lines etc. How is a call routed through the network? How does the system sort out STD codes?

Where does computing come in? Can I do some data processing with my terminal? Can I use it for telex, teletext, etc.? Can I use my PC and its word processor to set up telexes, electronic mail, etc.?

How can you make better use of your telephones – through BT and other suppliers?

What are the essential differences between Telex, voice, slow data and the various fast data services?

Will cellular telephone provide a reliable mobile data terminal?

What are System X and digital data transfer?

What is ISDN and when is it coming?

What is teletext? We know it is an international agreement and will replace electronic mail, but is there a network in existence; how do we get into it?

There seem to be very recent developments whereby telex, slow data, teletext, fax, etc., will be mutually interfaceable. How? When?

LANs. What are they? What benefits? What is available? What kind do I want? How do I work out what hardware/costs, etc. Can I get into other people's?

Viewdata/videotex seem interesting. What is the difference? Can I get into other people's? How? Could I set up my own?

What are all these protocols and standards, especially V.24, X.25, OSI.

Can I have a list of buzzwords and brief explanations?

How have things gone over the last 5–10 years and where are they going because, now I am interested, I want to plan for the future?

---

doldrum period when 'nobody would tell me anything', Nigel Stewart for help with the proof-checking, and my friend and colleague Osie Pereira (Head of Science and Technology at Slough College), whose patience in explaining some of the trickier concepts at idiot level for me cannot be valued.

Without all this help, cooperation and guidance, I could never have sorted the book out.

Incidentally, you will be very agreeably surprised by the presentation and content of much of the trade literature. Companies like DEC, Rank-Xerox and Plessey produce beautiful and very readable material.

Finally, as an overview of this book, you might be interested to see the notes I made in the very first hour's work I did for the book. These are reproduced above.

M. Gandoff

Chapter One

Communication and computers

The development of voice and data transmission

Let us start with a point raised in the Preface, i.e. the difference between telecommunications and data communications. When long-distance communication first started in the 19th Century, the name used was *telegraph*, 'writing at a distance' (from the Greek words). Then came the word *telephone*, 'speaking at a distance'.Since then we have seen *telex*, *teletext*, *teletex* and so on.

Voice and telex transmission technology have developed since then and were in very widespread use by the end of the Second World War. At this time, the computer started to appear and was well established by the late 1950s.

No apology is made for the fact that much of this chapter is concerned with computing and data processing as a lead-in to data communications, rather than with telecommunications. Almost all data transmission is controlled by computers these days and you need to have a feel for data processing in order to appreciate what is happening and how the various services and facilities fit together and can be accessed.

Before data communications became possible, computer processing had to be from a main computer within the building where processing was required (often referred to as *in-house* processing).This meant that users, those that needed their data processed and results generated, had to take their source documents (invoices, time-sheets, production schedules etc.) to a data preparation bureau who would produce punched cards or paper tape containing the 'source data' in

a coded form. Then these would be taken into the computer room and fed into the input device (card- or paper-tape reader) under the control of the relevant program. Only then could files be updated and printed reports produced. By the early 1960s, both manufacturers and users of computers were becoming painfully aware that there had to be a better way to gain access to computer power than this, the only route available at the time.

This not very 'user-friendly' performance was given the name *batch processing*, mainly to indicate that data originated in a batch, as far as the computer was concerned. The computer processes (programs) required would be carried out so that the batch retained its identity right through the data processing cycle. It was becoming increasingly obvious that a method was needed that would provide computer processing (access to the computer's processing power and data storage capacity) *at the point of service*, i.e. where data or the request for computer processing originates rather than at a central or host computer. For example, in a warehouse, it would be nice to be able to enter stock deliveries and at the same time have the computer record them so that other users could request up-to-date stock reports. Similarly, an engineer on a customer's premises might need to carry out a complex stress calculation using his own company's computer and the same might apply to a marketing man wanting to do a fancy sales forecast or delivery date calculation.

A manager could receive a lot of decision support from a computer if he had access to a range of databases containing financial and business information, to management science and statistical software. Obviously, only large companies with adequate resources and know-how could provide this kind of facility in-house.

Remember that the micro with any respectable software has only been around since about 1980 and in the early to mid-60s, computers were either fairly small and unsophisticated or large and expensive. So until recently, there was always the problem that data processing could only be carried out in batch mode.

Public voice telephony on the other hand really presented no such problem apart from the general difficulties of trying to reduce mis-routings and background noise and generally increase the efficiency of the telex and public telephone systems with the increasing level of traffic intensity.

Obviously, both the military and business uses of computers were being rapidly recognised and developed and it was only a matter of time before electronics and telecommunications equipment suppliers started to make extensive use of solid-state circuitry leading up to the now very widespread application of microprocessors. Similarly, computer manufacturers were able to produce the software that could drive data and interconnect the central computer and other computers or *terminal* devices such as VDU/keyboards.

It is not really surprising that things have only just started in a very big way. There are quite a few problems associated with communications of all kinds, related to the different switchboards and exchanges in use, the different terminals used, the different computers that back up much of the processing and the fact that we are connecting public and private services, in the UK and abroad. (In recent years, there has been implemented an international network for data transmission which links many different countries.)

For telex, the ASR33 teletype, first produced by Bell in the USA, had been in use for quite a while. This was basically a keyboard/slow printer that could be linked into the telex system both for sending and receiving and it seemed obvious that the teletype provided a means of connecting to a computer, *at a distance from it.* Equally obvious was the use of the telephone system as the

communication medium, the telex lines being a little too slow.

To save connect-time, i.e. to avoid having to enter the wording of the telex while connected, the teletype was fitted with a paper tape read/punch and as keys were pressed, as well as being transferred to the printer paper-roll, they were copied onto the paper tape in ASCII code (coming up later). Once the text had been completed, a telex connection could be made and the actual message entered from the tape through the tape reader. (In fact, paper tape has been employed in this way right up to today, although its use is dying out since 'intelligent' telex machines are often equipped with magnetic disk or some other method of storing messages before transmission, such as extra magnetic memory.) The ASR33 remained in use as an unintelligent computer terminal (no data processing power of its own) for quite a long time, the user preparing data and programs on the off-line paper tape punch and then feeding them in once connection had been made to the computer.

So, once industry had cracked the problem of converting computer-acceptable data, such as from the teletype, into a form that could be transmitted through the telephone system, and then getting it back into an acceptable form for the computer itself, *on-line* or *remote* computer processing became possible.

Of course, the software needed to be rethought, both to be able to handle processing for a number of remote users (not much point if just one batch

Figure 1.1 *The ASR33 teletype*

user at a time can be serviced) and to sort out the communications problems, such as who gets service next, error detection and recovery, collecting statistics of facility usage (for control and future planning), etc.

In the mean time, the Post Office had not been idle and could see the potential for data transmission as well as the large increase in the use of voice telephone and was already planning for what was to become the completely new, British Telecom *System X*, the computer-controlled public switched voice and data networks. The new technology was also implemented for the eventual introduction of *ISDN*, the Integrated Services Digital Network, which is intended once the full network has been implemented, to be used for the transmission of all kinds of data.

Before going on to review computer applications we must say a little more on the concept of a 'network'. The term is probably the most widely used word in communications and basically means a set-up of telephones and/or terminals, with a connection path or transmission medium and a mechanism for controlling the interconnection or switching of calls. That supplied by British Telecom for speech and low-speed data is called the *PSTN (Public Switched Telephone Network)*. The more recent extension to this is the *Packet Switched Network* called *Switchstream*. Telex is handled by a different network although there has been increasing interest and usage of *teletex*, a standard approach to electronic mail which may eventually replace telex.

Since 1981, after the Telecommunications Act made it possible for companies other than British Telecom to offer public network services, Mercury Communications Ltd was formed by Cable & Wireless (a British company that supplies a wide range of telecommunications equipment and services overseas), British Petroleum and Barclays Merchant Bank. It is now wholly owned by Cable & Wireless.

Mercury are now able to supply a network which complements and can provide an alternative to that of BT. A number of companies, such as Plessey and GEC, provide complete networks for private use, such as might be the case on a factory site or an even wider area using microwave or laser equipment to transmit data between points or complete networks that link in with the BT network. Yet other companies supply *VANs (Value Added Networks)* in which existing BT and Mercury networks are used and extra facilities are provided, e.g. electronic mail and home banking.

Networks that cover an area larger than a single building, whether for public or private use, are given the name *Wide Area Network (WAN)*.

On a completely different level, for several years now, many companies have been putting together and supplying their version of a *Local Area Network (LAN)*. The main difference between a WAN and an LAN is that the WAN can cover an area from many metres to the whole world, while LANs are usually aimed at rooms, floors or possibly whole buildings. In addition, the LAN is usually geared to a smallish number of users, with data transmission not relying on telephony techniques. An LAN will involve between, say, four and dozens of users, while a WAN could link thousands of distant users and may be based on speech or data transmission techniques.

A review of data processing

In order to appreciate the facilities available to the remote computer user, it is necessary to do a capsule review of what data processing is about. So let us start with the machinery and equipment.

Hardware

Computers themselves consist of three basic components: memory, the processor and peripherals.

Memory The electronic *memory* (these days usually called *RAM* – random access memory), is the electronic part of the computer into which data must be temporarily loaded so that it can be processed and into which programs must be loaded so that they can run. This is often backed up with a slightly different type of memory with the generic name *ROM* (read only memory) in which programs can be permanently stored for immediate use. The difference between RAM and ROM will come out when we talk about software.

The basic construction of any kind of memory is

a unit which can take on one of two values, conceptually *1* and *0*. The unit is called a *bit* which is a contraction of *BI*nary digi*T*.

A decimal number, such as 12345, is a shorthand for the sum:

$$(1 \times 10^4) + (2 \times 10^3) + (3 \times 10^2) + (4 \times 10^1) + (5 \times 10^0)$$

the digits (1,2,3,....) in each case representing how many particular powers of 10 in each position.

With a binary number, instead of powers of 10, we are dealing with powers of 2, so, for example, the decimal number 123 (or 123_{10}) can be written as

$$123_{10} = 0 + 64 + 32 + 16 + 8 + 0 + 2 + 1$$

and since $64 = 2^6$, $32 = 2^5$, etc., we can see this as

$$(0 \times 2^7) + (1 \times 2^6) + (1 \times 2^5) + (1 \times 2^4) + (1 \times 2^3) + (0 \times 2^2) + (1 \times 2^1) + (1 \times 2^0)$$

or

$$01111011_2$$

the *subscript* of 2 shows we are looking at a binary rather than a decimal number.

Now, memory is built up of groups of bits called *locations* usually 8 or 16 in number, so that each group can be individually referred to by an *address*, e.g. the first location is usually given address 0, the second 1 and so on. In order to hold numeric data in memory, we can use one or more locations, whereby each bit represents a power of 2. Held in, say, location 10 (16 bits), 123_{10} might look like this:

	0000000001111011	
9	10	11

If we want to store a number such as 12345, we can also hold it as a coded set of five individual numeric characters, using four bits to represent each digit whereby:

$$1 = 0 + 0 + 0 + 1 = 0001_2$$

$$4 = 0 + 1 + 0 + 0 = 0100_2$$

$$9 = 1 + 0 + 0 + 1 = 1001_2, \text{ etc.}$$

This is called *BCD* or *Binary Coded Decimal*, and is still used by some microprocessors for handling decimal data. Its disadvantage is that it cannot be used to represent non-numeric characters such as letters, punctuation, etc.

For many years, Telex transmission has used the *Baudot* code, also called the International Telegraph Alphabet no. 2 (ITA2), which is based on five bits. (Incidentally, Baudot, a very early worker in telecommunications has also given his name to the 'baud', a unit related to data transmission speeds.)

An improvement on BCD was to use the same BCD bits extended with two more bits making a six-bit code called *EBCD* (*Expanded Binary Coded Decimal*) code.

In this, the first two bits indicate the range of the characters (A–I, J–R, /S–Z, 0–9) and the BCD bits specify where in the range. So:

Bits 1–2	BCD bits	Range of characters
11	1–9	A–I
10	1–9	J–R
01	1–9	/S–Z
00	0–9	0–9

Hence,

110011 = C (3rd in the range A–I)
011001 = R (9th in the range J–R)
100010 = / (1st in the range /S–Z)
001000 = 8 (8th in the range 0–9)

In addition, 110000 is a ? and 100000 is a !.

Other characters can be represented by other combinations of bits, where the lower four are greater than 9 but even with these, since there are only six bits, there are only 64 different combinations – we have used 38 already and we have not even enough left for lower-case letters (a,b,c,...).

There is a further extension to EBCD used mainly by IBM, called *EBCD Interchange Code* or *EBCDIC*. It has a further two bits, making an eight-bit code. This allows for 256 combinations, of which about 100 are used to represent

Print character	ASCII code 7 bit	Print character	ASCII code 7 bit
space	010 0000	P	101 0000
!	0001	Q	0001
"	0010	R	0010
£	0011	S	0011
$	0100	T	0100
%	0101	U	0101
&	0110	V	0110
'	0111	W	0111
(	1001	X	1000
)	1000	Y	1001
*	1010	Z	1010
+	1011	[	1011
,	1100	\	1100
–	1101	]	1101
.	1110	↑	1110
/	1111	–	1111
1	011 0001	'	0000
2	0010	a	110 0001
3	0011	b	0010
4	0100	c	0011
5	0101	d	0100
6	0110	e	0101
7	0111	f	0110
8	1000	g	0111
9	1001	h	1000
:	1010	i	1001
;	1011	j	1010
<	1100	k	1011
=	1101	l	1100
>	1110	m	1101
?	1111	n	1110
@	100 0000	o	1111
A	0001	p	111 0000
B	0010	q	0001
C	0011	r	0010
D	0100	s	0011
E	0101	t	0100
F	0110	u	0101
G	0111	v	0110
H	1000	w	0111
I	1001	x	1000
J	1010	y	1001
K	1011	z	1010
L	1100	{	1011
M	1101	|	1100
N	1110	}	1101
O	1111	~	1110

Figure 1.2
ASCII characters

characters and the remainder are used for control signals, etc.

A more universal code as far as telecommunications is concerned, is based on seven bits and is called the ISO 7-bit code. This is slightly extended by one more bit (the parity bit), used for error checking in the *ASCII* or *A*merican *S*tandard *C*ode for *I*nformation *I*nterchange.

The arrangements of bits are not quite as logical as with EBCD as you can see from Figure 1.2.

The processor This is the bunch of electronics components and logic circuitry that carries out the actual processes needed to satisfy applications problem-solving and to control the interaction between all the hardware and the software. Some large computers are based on several or even many processors, depending on their purpose. For example, a typical minicomputer could have separate processors for dealing with communications, files/databases and the applications programs themselves. For microcomputers and some minicomputers, the processor is a single 'chip' or integrated electronics unit and is usually called a *microprocessor*, commonly used ones being the Zilog Z80, Intel 8080 and more recently the Motorola 68000 series of microprocessors.

Very large machines may have linked *processor arrays*, where many applications may be going on simultaneously in many processors and a recent innovation is the *transputer*, a completely new kind of processor which can carry out many processes at the same time. It is expected that they will be fitted in microcomputers in the very near future.

On a slightly different level, some sophisticated processors can actually handle instructions from different applications at the same time.

So, a typical processor is built to carry out certain electronic *operations* in response to coded signals called *instructions* (memory being needed to store these instructions prior to execution). Instruction codes like data codes, are based on zeros and ones and contain two basic parts. The first indicates to the processor, the operation to be executed (carried out) and is usually referred to as the *operation code* or *opcode*. The second part usually indicates the addresses of the memory locations in which the data to be acted on are stored.

Typical instructions in *machine code* as it is called might carry out operations such as:

MOV 100,200	Move a character from location 100 to location 200
ADD 100,500	Add the 16-bit binary number in locations 100–101 to another 16-bit binary number in locations 500–501
ADDI 100,500	Add the 16-bit binary equivalent of 100_{10} to the 16-

bit binary number held in locations 500–501

OI 100,X'00'	Set all the bits of location 100 to zeros.

The peripherals These are the input, output and storage (input/output) devices that feed and are fed by the processor.) Examples are:

Input devices:
VDU or electronic typewriter keyboard
Mouse
Light pen
Barcode reader

Output devices:
VDU (cathode-ray or plasma screen)
Printers (daisywheel, matrix, laser)

Input/output devices (or *storage* devices, or *backing store*):
Magnetic tape (reel or cassette)
Magnetic disk (floppy or hard, single or multiple surface, fixed or removable).

Figure 1.3 *A mouse in use*

Figure 1.4 *A daisywheel printer*

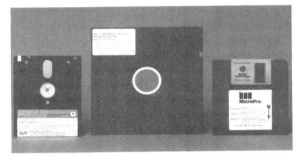

Figure 1.5 *Magnetic diskettes*

Sometimes data can actually be generated in memory (perhaps as the result of a calculation), say in binary or in ASCII, but more often it will be recorded or stored outside. The processor needs to be able to fetch this data into memory for processing and to send out the results afterwards. Now, data is usually created in a human-readable form (as the 'source document' mentioned earlier). In order to get it into the memory, there needs to be an interface that can convert the source data into a form that can be loaded into memory and this is where the input devices come in.

However, codes such as ASCII have application for input and output as well as in memory. With paper tape, which has been around for many years, characters are represented across the width of the tape with small circular punched holes, using the convention that 'hole' = 1 and 'no hole' = 0. In a similar way, data is stored on magnetic disk or tape as a series of magnetic spots which like memory can be set to 0 or 1 and hence groupings can represent ASCII or EBCDIC characters.

Although not strictly definable in terms of hardware alone, we should try to distinguish between mainframe, *mini-* and *microcomputers*. This distinction is becoming more and more blurred these days and perhaps the only guide, crude as it is, is what the manufacturer calls his product.

At one time, the size of the memory, speed of the processor and the capacity and data transfer speeds of the peripherals were some guide. For example, at one time, a mainframe processor might have a memory size of four million characters, carry out about ten million instructions a second, have disk drives that could read/write perhaps 250,000 characters per second and be fitted with a printer that could handle up to 1500 lines per minute.

However, at the other end, typical micros are now being supplied with very large memories (up to 16 million locations), disk drives are available that can access hundreds of millions of characters and 'small' printers are getting faster. It is becoming very difficult to say exactly where a particular machine fits in. However, the micro is still the smallest in physical size: most of them will fit on a desktop and are portable. (This book was word-processed on one that weighs a very heavy 40 kilos! A more typical machine weighs much less.)

The mini usually looks rather larger and is certainly not portable. Its disk size and capacity will possibly not be much better than the best micros but its processor is likely to be very much more powerful and more able to support *multi-programming*, where a number of programs can apparently be undergoing execution at the same time, not only within the computer room itself but, by using the technique of *time-sharing*, the programs can be run by many remote users.

Micros have been able to do this, but until very recently, only to a limited extent. Some of the more recent microprocessors, such as the Motorola 68000 series, are used in micros and in minis and they would appear to be capable of much more extension. The extent to which they are employed will depend on what the user wants rather than the limitations of the chip.

The software that is provided with micros and

minis has been designed so that the 'user', or non-computer-experienced staff, can handle it without expert assistance.

The mainframe is generally installed in its own 'computer room' and requires operators to cope with it. Similarly, the software is not so 'user-friendly' and in effect, the operators need to communicate the user's requirements for operational program running to the computer software.

Software

The processor carries out *operations* such as for arithmetic, data movement and manipulation and activates the peripherals with commands, these being in response to *instructions*.

A *program* is a set of instructions to the processor presented in the order required to make it carry out a particular task. The form of these instructions is the binary-based, machine code instructions already mentioned, each one specifying the operation to be carried out and the *location* of the data to be acted on, either in memory or at the device from which it will be accessed.

Programming languages have been developed since the beginning of computers, because writing programs in machine code is generally very difficult.

The languages act as an environment in which humans can set out the logical steps of a program in a form that is familiar to them. Where necessary they can be like the form of the language used for the problem application area the program is being written to solve. Hence we have languages for circuit design, banking, mathematics, management science, etc., as well as fairly general-purpose languages such as Fortran, Basic, Pascal, C and, of course, Cobol, the business-oriented language which has been in constant use since about 1957.

The program once written as a set of language statements is converted or translated into machine code by a translator program, variously called an *assembler*, *compiler* or *interpreter* depending on how it actually carries out the translation and the kind of programming language it has been written to cope with.

After translation, the program is usually stored on disk and can be loaded into memory for execution at any time. However, there are some applications where it is necessary to have the program in a more available form – it takes a certain amount of time to read programs from disk and even if the program is kept in memory for a whole session, RAM is generally not able to hold machine code when the machine is switched off.

In addition, many different applications may need service at the same time and it is just not convenient to have to load from disk. For these applications, ROM in various forms is used. The word ROM is an acronym for *Read-only Memory* which means that once inserted into ROM, machine code instructions are held permanently. The ROM is purchased, pre-set from the supplier. When you purchase some add-ons to your micro (that will crop up later on in the book), you will often find that they will include some ROM and are usually called *cards*. The dealer can say that a 'memory-extension card' can be provided.

There are variations on the basic ROM. *PROM* (programmeable ROM) is obtained 'blank' and can be set, once-off, to whatever the user wants. *EPROM* can be set like PROM, but the user can blank it off and re-set it when necessary. There is a variation on this whereby the ROM can be 'opened up' for re-use in response to a special signal. This is useful in large networks where terminals are pre-set to carry out certain processing. Where a modification is needed, instead of providing an upgraded card, the main computer can alter all the terminal ROMs, without the user being aware of the change.

Programs stored in ROM are often referred to as *firmware* and incidentally, some people may use the term 'software' erroneously in the sense that it is disk-based rather than in ROM. It is all software really.

ROM is employed essentially either when the disk-load time is prohibitive, or when an application is pre-set in the sense that the range of software needed is limited and can be held 'on-line' on a permanent basis.

Typically, parts of the operating system (the software that controls the use of a computer) are held on ROM, so that with many microcomputers for example, as soon as the machine is switched on, it starts the initial set-up dialogue with you, such as asking for time and date. The instructions for this are in ROM and this software may actually cause the disk-based software to be loaded when needed.

Perhaps of more relevance to communications is the increasing trend towards *intelligent* hardware. The term 'intelligence' just means the ability to carry out some processing. Hence, a non-intelligent or *dumb* terminal is one that does little more than enable data and commands to be sent and received. An intelligent terminal will, at least, be able to format data, carry out error-detection and perhaps do some data vetting. Often, complete microcomputers are used as intelligent terminals and can take over some of the application-program running, thus leaving the central processor at the other end free to handle the network and service users that do not have intelligent terminals.

In essence, whenever and wherever a pre-set of operations is needed, a processor/ROM combination can be employed. The hardware that handles network management, accounting and control is highly intelligent even though it does not load software from disk and the same applies to the modems, concentrators, multiplexers, etc., that form part of most networks.

A classification of software

It may seem a little strange to be talking about software in a book about communications, but to repeat what we have already said, the bulk of data transmission, which forms a very large part of communication, is either intended to initiate computer processing or is controlled by computers. In order to appreciate the significance of a computer at the other end of a telephone line, it is important to be aware of computer applications and, in particular, the programs that are needed to support them. In addition, the 'dialogue' or conversation between user and computer is intimately related to software and is a very important part of terminal-oriented data processing.

There is such a wide range of software that we almost despair to try and organise a description. But if we start with a broad classification of 'communications services' (Figure 1.6). We can then look at a similar classification of software, as shown in Figure 1.7.

Voice is obvious, but is now backed up with extremely sophisticated technology permitting automatic and semi-automatic dialling, call-charging, exchange routing, conference calls, etc.

Telex is still probably the second most important communications method for busines although not for much longer. Until fairly recently, telex required special terminals which were specifically designed to hook into the telex network. Now there is increasing use of specially adapted microcomputers whose software backup gives advanced telex facilities such as deferred transmission, charging, word-processed text, etc. Some of these systems allow for telex to be sent and received while the computer is carrying out other operations.

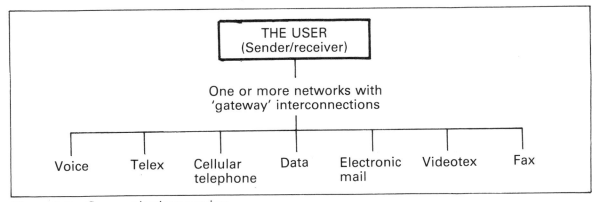

Figure 1.6 *Communications services*

Cellular telephone has exploded in the last couple of years as a convenient means of mobile voice communication in which a radio-telephone links into the telephone network via a series of transmitters covering a number of 'cells', i.e. geographical areas. Quite recently, there has been talk of data transmission as well as voice, but error-detection and handling is still a problem since it involves transmission along wires and through the air.

Data transmission as distinct from voice or text implies specific communication with a computer for immediate or later data processing. The computer and its processing ability can be owned by the user's company or could be offered on a contract basis.

Electronic mail is technically the same as data transmission but when referring to text, we usually mean electronic mail, where messages, letters, reports etc., can be sent as an alternative to telex, through public, and wide- and local-area private networks. The facility could be a simple message or bulletin board in which users can leave and pick up electronic messages or it could be a rather more complex system such as Mercury's *Easylink* provided by Mercury or British Telecom *Gold*, which as well as giving a messaging system, also provides a word processor and other electronic office facilities.

Viewdata which is also called Videotex, is a general term for a combination of text and data processing which broadly covers the ability to obtain information in a highly readable form from large stores of data provided either for public use, such as by Prestel, Dialog and Pergamon, or the private, corporate videotex networks now being used within many companies implemented by DEC, Xerox, Mars and others.

Thomas Cook, the travel agents, have recently had a videotex network for all their offices installed by Microscope Ltd. This makes it possible for booking clerks to obtain up-to-date information on holiday availability, airlines, currency, etc.

Fax (facsimile transmission) is a relatively new technique which uses a technology something like photocopying, to convert electronically the contents of whole documents such as diagrams and drawings, and transmit them to a distant receiving machine.

What has become very exciting is the fact that users now expect to access several of these facilities from the same terminal and since many of them have microcomputers as terminals, the fitting of some extra hardware and some extra software means that a user can obtain information from a videotex supplier, work on it with a word processor provided by another network and then telex or fax the result, perhaps to an associate company abroad. The linkage between the different networks is made possible by what are called *gateways* which are basically hardware and software links that provide access to one gateway from another. (In the next chapters, you will see examples of gateways for telex-to-PSS, etc.)

Systems software

This can be loosely divided into software that helps with the writing and testing of applications software. Someone has to write and test programs before they can be used live and other people produce software to assist in this. We have already mentioned translator programs such as for Cobol, C, etc., and, in addition, much software is available for facilitating the job of producing operational programs.

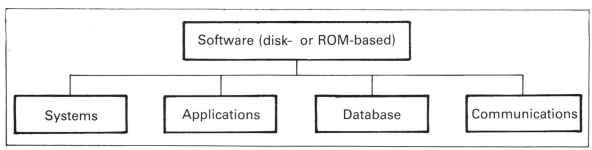

Figure 1.7 *Main software applications*

At one time there was a clear distinction between the 'user' as the person who wanted the computer to be of service, the programmer who produced the programs and the operator who was responsible for making sure that the programs run operationally to produce the results needed by the user.

Now, the typical microcomputer user has available a wide range of applications software, a lot of which can be adapted or modified to meet special requirements, so although programming language translators are available to enable the user to write his own programs, there has not been a major trend towards this. People prefer to buy complete, pre-written programs and use them 'as-is', or modify them to their needs.

Of more interest importance to the user, as well as the professional, is the software that can assist in the efficient running of applications software. It is usually identified with the name *operating systems*, although these days the phrase 'user/computer environment' might be more appropriate. With this software, we are trying to make it as easy as possible for programs to be run, either by operators in a computer room, or by terminal users. Remember, although you think you are using a *facility* called a spreadsheet or word processor, in practice, a *program* has to be run to carry out the activity and the computer needs to be told which program to run and so on. The user is not concerned with programs, but needs a way of telling the machine what to do and this is why good dialogue facilities must be written into the programs. These help the user to:

- Select the functions he wants to carry out, i.e. the programs to be run.

- Enter instructions and data as easily as possible, with the minimum possibility of error.

Figure 1.8 *A typical menu*

- Report on errors made showing what to do about them.

- Display information requested in an acceptable, understandable way.

Menu-driven systems Until fairly recently, system choices were usually made by selecting from a 'menu', a list of VDU screen options as shown in Figure 1.8.

Each option represents a function to the user, but the software behind this could include several programs. All the user needs is a way to indicate his choice. Traditionally, you enter the number of your choice through the keyboard or used the 'cursor-control' keys, $\rightarrow$, $\leftarrow$, $\downarrow$ or $\uparrow$, to align the cursor over the choice. Sometimes the menu is displayed across the top of the screen as shown in Figure 1.9.

The light-pen has been used to assist in selecting from menus. It is like a pen and with it, it is possible for the hardware to detect where it has been touched on the VDU screen. More recently, the 'mouse' has become very common. This is a hand-controlled device which is moved over the surface of the desk making the cursor move correspondingly. When it is in the correct position, a 'fire' button can be pressed.

Figure 1.9 *A spreadsheet 'top line' menu*

On the basis that 'a picture is worth a thousand words', a lot of software these days makes extensive use of 'icons', i.e. shapes on the screen that represent functions. For example a filing cabinet to denote saving a file just created or amended, a 'paper' symbol to request a print-out or a waste-bin to show that a file is to be scrapped. It is not only convenience that dictates the design of this dialogue. In spite of computers being in pretty widespread use for at least 30 years (J. Lyons first used a commercial computer in England in the mid-1950s), many users are still put off by keyboards and seem to find the mouse more 'user-friendly'.

This has been extended very much in *CAD (computer-aided design)*. Packages are available with which the 'menu' is a wide range of shapes representing engineering, electronic or building components and, by skilful use of the mouse, it is possible to produce drawings and designs very quickly. The end product can then be reproduced on a colour plotter.

The examples of dialogue shown could be produced by the applications programs they relate to, but there are several pieces of software available for micros that act as an environment or interface between the applications programs and user, such as *GEM* by Digital Research (see Figure 1.10) and *WINDOWS* by Microsoft.

An operating system does more than produce pretty VDU screens:

● It will control the use of the processor and other hardware.

● It will detect, report on and sometimes correct many kinds of error.

● It will make possible the automatic running of jobs (programs) and make the best use of facilities available (memory, time, disk space, etc.).

● It will link together applications programs, communications and database software.

● It will time and account for the use of the computer so that users can be charged.

● It provides a means of communication between user, programmer, operator and other computer systems.

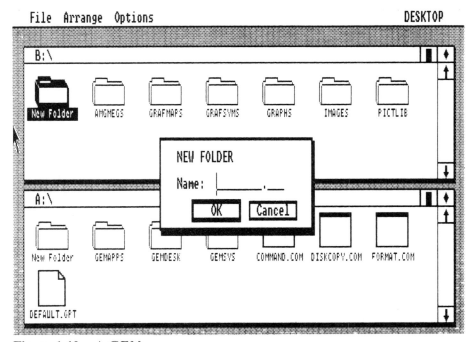

Figure 1.10 *A GEM screen*

Modes of computer usage

Before looking at specific applications, it is appropriate to consider how a user gets access to the computer.

In-house batch processing As far as mainframe and minicomputers are concerned, programs are run *in-house*. Computer operators in conjunction with the operating system arrange that the relevant programs are run according to a schedule.

The input data is prepared for the computer in advance of the job and the printed output is returned to the user or sent off, e.g. invoices to customers. The user never sees the computer and certainly never enters data through a terminal.

Micros can be similar in that programs are initiated by the operator, but due to the much smaller data volumes, input is often through the keyboard (with a VDU screen to make it easy). But the big difference is that the user is often his own operator and data entry clerk. The typical micro user buys a computer and a set of programs and either employs or re-employs a clerk to 'run' the computer. After suitable training, this person should be able to run programs, enter the necessary data, look after files and produce the necessary output documents and management reports.

The advantages of using a 'dedicated' computer, irrespective of its size, are obvious. You have control over it and can use it when and for what you want (perhaps even selling off some of its surplus capacity). The disadvantages are that you must maintain both the hardware and software and you will probably need to have much of your software written for you.

On-line processing When jobs are run in batch mode, there is an inevitable delay, for two reasons: By definition, data volumes are large (because documents have been batched up) and as a result, program run-times are often minutes or hours rather than seconds. Secondly because many jobs are run, computer usage must be scheduled. Low priority job A may be requested early in the day and may actually start running. Soon afterwards, higher priority job B starts and takes up the processor and memory space. This means that job A is delayed. Even if job A has

actually finished, its printed output may have been 'spooled' to disk (to optimise hardware resources, program output is often temporarily placed on a disk file for later printing out) and although its 10 pages of output are ready to be printed, the 100 pages from job B must come first.

As soon as there are a number of users wanting access to almost immediate computing power, batch processing becomes irrelevant and a different operating environment becomes necessary. These on-line users will want computing power for any of the following reasons:

Enquiry This is fast becoming the largest single mode of computer usage. Files or data stores are an integral part of any system and are created either as a result of earlier processing, such as a customer file which records invoice values, payments, credit notes, sales value this year, etc., or specifically, by 'information providers', so that they can provide the basis for general information retrieval. This latter is usually called *Viewdata* or *Videotex*.

In the first case, a company will need to maintain its books on a computer in the same basic way that it does in a manual system. This will mean keeping a customer ledger and being able to enter or 'post' records of all transactions affecting an account. During a processing period, the computerised ledger will be used to produce invoices, credit notes, statements and so on. However, at any time, it might be useful to produce analyses, summaries or exception reports to make the job of management easier.

In a batch processing system, it may be a regular procedure at week- or month-end to produce profitability reports on products and processes. A summary of sales and costs by stock item or item group could provide useful information about which lines are profitable or where there are production inefficiencies.

Similarly a summary of customer sales by customer type within a sales area can help a manager decide whether to put in more or less sales effort or arrange promotions in a particular sales area or for a particular type of customer. An exception report might be produced in which the details are printed out for any customer who fails to satisfy certain criteria, such as 'no payment for three months', 'outstanding balance greater than £1000' and so on.

As we introduce increasingly more of an on-line aspect, information such as this must be provided when requested, not just at month-end, so we are back to dialogue again. The user may want to specify a number of options relating to the information to be displayed. The software to allow this and to provide the nice screens is the main part of the 'database' and 'communications' software.

Viewdata is the generic name given to a system that allows for an ordered access to a structured set of data in such a way that enquiries can be serviced to generate useful, readable information and has two main sources. The first is for public or wide, general use and is said to be supplied by *information providers*. The second is often called 'private' or corporate viewdata and is an extension of the more traditional, on-line facilities provided by a company for its staff. When such a system is to be implemented, a private communications network is often needed. The information providers, viewdata, and its hardware requirements are discussed later on.

Low volume data entry (such as recording stock issues and customer payments at a supermarket checkout or holiday reservations at a travel agent). To support this, the software will need to validate (check out) the entered data and help the user to get it right. It will then need to be able to handle the transmission of the data to the correct central, or *host*, computer by the most efficient route.

This data could be sent to a central computer and stored (batched) for later batch processing as is the case with the clearing banks. From cash points or by requesting the bank clerk, a statement request is passed to the computer for overnight production.

On-line data entry plus update is more likely. We want the computer to carry out processing so that as well as transmitting the data to the host, it will need to be used to update the various files or data stores that are the backbone of the application system. As stock issues are entered, it may be important for the stock file to be updated straight away, so that other users soon know of revised stock levels. This is particularly important in booking situations such as airlines, holidays, etc.

Obviously the quicker a booking (or cancellation) is processed, the quicker will all the terminal clerks be aware of the changed situation.

The cash points provided by the clearing banks are on-line terminals which are used for low volume data entry and update and for enquiry as well.

A similar facility using a Prestel TV or microcomputer as a terminal is available for shopping and banking, called *EFT (electronic funds transfer)* or more particularly *EFTPOS (EFT at point of service)*.

The Nottingham Building Society offers *Homelink*, an on-line service. So does the Bank of Scotland with its *HOBS (Home and Office Banking Service)*.

Figure 1.11 *Homebanking*

British Telecom will supply a Prestel adaptor for your TV so that you can access these services through Prestel, or you can use your micro fitted with a suitable modem to get at them directly.

Applications software

These are programs written to carry out jobs to assist in a particular application. There are a number of terms used in this area which need definition. The word 'system' can be used as follows:

Computer system The computer that is providing the processing facility. (For larger computers the specification could also include details of the operating system.)

Computerised system   This is a more widely embracing term and generally means the programs and manual procedures needed to carry out an application, such as payroll, sales, or production.

Program   A single set of instructions to the computer for a particular task. For example, a traditional computerised payroll system might include the following programs:

- Enter and validate employee master data and add new employees/changes to the master file.

- Copy the master file after the update.

- Enter and validate weekly time-sheets.

- Calculate tax and NIC and produce payslips, cash analyses and P45s for leavers.

- Produce sickness and holiday analyses.

- Produce overtime and idle time analyses by department, etc.

- Produce annual summaries and P60s

Package   This is one of the most over-used names in data processing. Its original application was to a system (programs and procedures) purchased from a supplier or written for you to your requirements. In fact, an 'off-the-shelf' package is one that you buy complete and either adjust to your needs, or change your methods to fit in with it. A 'tailored' package is developed specifically for you. A 'customised' package is one written for general purposes and then modified by the supplier to your requirements.

Software package   Probably the silliest term in data processing. A package is either a group of programs, in which case it should be called a 'system', or it is a single program in which case it could be called a 'program'. However, the word is probably the biggest 'buzzword' in computing today.

Software sources

There are four main ways to obtain programs for particular applications:

- Employ analysts to specify them and programmers to write and test them.

- Have a package written for you.

- Buy a package.

- Use a package or application generator (a special kind of program which allows for automatic generation of the programs you need for a particular application).

- Rent pre-written programs.

This latter is actually more interesting than it first looks. It could be just paying a rental charge to make use of programs, the charge presumably including an updating service as improvements are made and mistakes ironed out. But the on-line user has a different route to software. Companies similar to the information providers already mentioned, and in some cases the same firms, supply the facility whereby users can run programs in a wide range of application areas from their terminals, using similar dialogue to that used in information access. This is particularly well supported in technical and engineering applications by companies like GEISCO and Boeing Computer Services.

Software applications

These are as varied as the people that use computers, but we can look at a broad classification:

Bookkeeping, accounts and finance   The recording of business and monetary transactions in order to produce final accounts and statutory reports (VAT, PAYE, etc.) and at the same time monitor the financial progress of the organisation (cash flows, profitability, etc).

Costing/budgeting   To set up budgets for products, processes, etc., and then by entering costs, to be able to control processing, manufacturing, etc., by comparing the budgets against actual figures. This could be linked with accounting software from which the cost figures might be extracted.

Resources control This could link with accounting and costing systems and includes control of stock, labour, vehicles and other resources.

Paper handling Strictly speaking a function rather than an application. There are areas in modern business where the sheer volume of paper handling requires the use of powerful computers. Imagine having a file of a million names and addresses and wanting to select a tenth of these according to many different criteria and then print an address label or even a personalised letter for each selection.

Management science/operational research This very broad subject overlaps many business areas since it is largely concerned with optimisation – making the best use of resources available. It includes many tools like 'network analysis' for project control, 'linear programming' for a range of problems involving the best use of materials and other resources, 'simulation' where the computer can be used to 'act out' business and technical processes under varying conditions to see what is the best policy, etc., and 'forecasting' which as its name suggests, requires the computer to predict values into the future.

Technical/engineering This has become very large in recent years and has become increasingly more linked to the user. The term 'CAD/CAM' (Computer-aided Design/Computer-aided Manufacturing) is popular and can be illustrated by an engineering designer using a computer to design a tool or some other product. The computer might then print final drawings, produce a bill of materials, estimate the labour requirements and generate a schedule. At the same time, it could generate a set of instructions and directives to automatic machines in order to automate the production/manufacturing process.

Electronic office Not one application, but really a whole group of individual programs that carry out very useful functions. These include word processing, spreadsheet (a convenience calculating program that in particular, enables columnar data to be manipulated), 'graphics' presentation, 'database', electronic mail and electronic diary, teletext and others.

Data/information access Data and the information that can be extracted from it are probably the most important resources owned by a company these days and there is a growing trend towards making information available to whoever wants or needs it, in the form that they want, whenever they want. This is obviously possible only within a heavily computerised framework geared to a database.

Database software (database management systems)

Every computer application needs access to stored data and it is now true to say that being able to organise data for easy and meaningful retrieval is an application in its own right. In the earlier days of computing and with the smaller applications today, systems were built around files. Data would be entered regularly after validation and summarised or stored directly onto files (usually on magnetic disk) for later reporting. Within a particular organisation, the overall company computerised system would have been developed as a set of sub-systems, perhaps at different times, by different people, each sub-system having its own files. Since all the sub-systems involve company business, there is bound to be some degree of data duplication. For example, the payroll files could contain, amongst other data, employee number/department/grade/salary/tax to date, etc.

The production handling system, in order to determine bonuses, might need employee number/ department/ grade/ contracts-worked-on, etc. Similarly, the staff administration/personnel system might need much of this data.

The result is a series of systems, each with self-contained files, the only links between them being through documents or specially produced interface programs/files. With different files being used for different purposes and being updated by different processes at different times, it is almost impossible to ensure that the files are consistent. If personnel have to produce a profile for an employee who is to be considered for promotion, income details will be needed and the personnel system can only obtain these from the payroll system, either as a report (or VDU screen) or possibly as a file especially produced by a payroll program to be accepted as update data by a special personnel program. If payroll has not been run, the profile will be out of date. If production has not been run, payroll cannot pay bonuses, etc.

By the late 1960s, hardware technology had improved to the extent that magnetic disks were becoming bigger, faster and more reliable. Computer memories were being made much larger and processors were becoming more powerful (shorter instruction times). At the same time, software technology was improving, particularly in relation to text and data manipulation. Since then the need for more powerful and more efficient operating systems, language translators, text editors and word processors spurred on the development of techniques for very efficient data handling and data storage which provided a sound basis for databases.

From a simplistic view, we can say that a database provides structured access to structured data largely independent of the application needing the access. Data is stored in a complicated, integrated structure and the 'database management' software makes it possible to get at the data either from a programming language in order to write applications programs or from the user himself, by means of what is called an 'enquiry' language. Generally, a piece of data is only stored once in the database which leads to a much more consistent set of data. In addition, once a change is made, all database users will be aware of it.

With information processing on a large scale, whether making use of a public system like Prestel, or one from an information producer like Dialog, Datastream, Extel, etc., the requirements are the same: a large mass of data, with many users who at any time may want to get at it for different reasons, from different aspects. The only possibility is a database.

Now, the subject of well designed dialogue becomes extremely important. Prestel as provided by British Telecom and similar systems throughout the world present 'pages' of information in roughly the same style with access via menus or page numbers. The general style tends to be followed by private information producers and when companies set up their own viewdata systems for use within the company. This subject is covered in a later chapter.

On a much smaller scale, the micro user has available a number of packages called 'databases' which by a combination of some file and some database techniques, make it possible to set up a small, simple database and to organise data processing and enquiry systems from it. These packages include Lotus 1–2–3, Symphony, Xchange, Open Access, and Jazz. Any of the following might be included in the package with data being accessible from and transferable between all of them: Database, Spreadsheet, Graphics, Word processor. Once the database has been set up, the spreadsheet can be used on the data to carry out calculations and produce summaries, etc. Additional processes may be carried out using the 'data manipulation language' provided within the database software. In its more sophisticated forms this can be almost like a programming language and allows the advanced user to extend the range of the spreadsheet and to increase access to the data beyond the data enquiry language.

A data enquiry language is often provided to make data access easy for the user. It is usually based on a series of statements with keywords which allow the system to produce reports and VDU screens showing summaries and analyses from different aspects in a variety of sequences.

Once processing has been carried out, the word processor can wrap the figures up into a report or letter and the graphics facility can be used to produce data in a visual form such as graphs, bar charts, etc.

Communications software

This represents all the programs whether at the host computer, the terminal or in the ROM of the intermediate network hardware, that handle such areas as the following:

- Controlling a whole switched or digital network including routing, priority handling, flow control, etc.

- Linking an applications host computer to terminals or other networks.

- Linking devices to devices.

- Handling and recovering from errors and hardware faults and failure, and rescheduling times and routes if necessary.

- Collecting statistics for monitoring, control and accounting.

- Controlling an LAN and linking it to other LANs and external services.

- Protocol conversion, to enable the interconnection of networks, devices and users which are based on different standards at the various levels concerned.

In the next chapter we can look at some of the theory behind communications.

Chapter Two

Transmission technology and techniques

Introduction

In the first chapter we were entirely concerned with the computer and its uses and usage. Now we need to look at the processes involved with the transmission of voice and data. As we will see shortly, telephone lines are not very good for sending data, except at low speeds, and even then, line noise and signal distortion can ruin a message. But, as telecommunications have advanced, so has data communications. The increasing level of voice and data traffic, together with technical advances such as the use of optical fibre, have lead to very sophisticated networks, in which voice and data can travel along the same paths at the same time. The teletype terminal has been replaced by much more sophisticated hardware such as the microcomputer, and now it is possible to have a terminal or grouping of terminals (*cluster*), with the facility to talk to other terminals, send or receive voice, telex, viewdata or fax and connected through the same highways to computers that can service all of these remote data processing requirements.

This situation has been made possible because people concerned have tried, by means of stand-ard usage and agreements, to ensure that a common approach has been adopted and a high degree of compatibility obtained between different hardware manufacturers, suppliers of computing power and other equipment and software producers. This does not mean that everybody has to make the same product, but rather that a supplier of a product or service must be able to guarantee that his facility can *interface* in a standard way. The word *protocols* is thrown about and it is important to know something about

them since they represent sets of rules that apply to the various levels of conversation, whether voice or the various forms of data transmission.

Since communication is international, protocols and standards often need international agreement and acceptance and many organisations have contributed to them or laid them down, such as ISO (International Standards Organisation), British Telecom, ANSI (American Standards In-stitute), ECMA (European Computer Man-ufacturers Association), IEEE (American Institute of Electronic and Electrical Engineers) and the CCITT (Comite Consultatif International de Telegraphie et Telephonie).

In addition, large computer manufacturers such as IBM, Xerox, Honeywell and DEC, as well as the larger manufacturers of telecommunications hardware, have laid down their own standards, some of which have been adopted by the world at large, either to make use of their kit or to interface with it. Other companies have developed hardware or software for *protocol conversion* which allows a user of one type to link in to services provided under a different set of standards. The subject of protocols is extremely wide and complicated, but we will try to give some ideas later on in an appendix.

As we have said, from the same terminal, a sophisticated user may want access to computers and what they can provide in data processing and information/data and in addition, to telex, voice, fax, other local users, Prestel and so on. These have become so intricately interwoven that it is now pretty safe to say that there is no longer any need to have 'telecommunications' and 'data communications' as separate terms; 'telecommunications' can now be used for voice and/or data. This 'getting together' of voice and data is usually called *convergence*.

Whatever names we use, the fact remains that looking at the global picture of modern communication, we can see that it is expanding at an explosive rate because it is driven by the technology behind it which is expanding at the same rate. In other words, the development of the hardware and its ready availability means that other manufacturers and service suppliers start thinking of uses for it and build it into their products.

While it is completely impossible for anyone to keep up to date, if you have an interest in modern methods of communication, you must have some familiarity with the concepts and techniques that are presented in this and other chapters in order to appreciate their scope, where and how they can be of use and how to go about selecting them.

So we must try and lay down a theoretical background/basis, hopefully simplified, and we should really start with the difference between data and information and consider exactly what is meant by communication.

Communication concepts and techniques

For communication, we must have a sender, a receiver and a link between them. With most of the techniques covered in the book there will be a certain minimum distance between sender and receiver, usually more than, say, 5 metres. So we are not concerned with data transfer between the mainframe or minicomputer and its associated peripheral devices – those directly connected to it, although we will be concerned with devices in a LAN (local area network) which could be less than 5 metres apart.

Regardless of what means we use for the transmission, essentially we are trying to communicate, implying the transfer of *information*: the sender is *informing* the receiver and hence, providing him with knowledge he did not already have. We must touch on the rather subtle distinction between information and data because they are not the same. In essence, information must inform, while data is the means by which the concepts we are passing are quantified or qualified, i.e. given some value. The receiver may well be able to receive data, but unless there is a pre-determined set of rules, or it contains its own mechanism for interpretation, it may not be possible to extract the *meaning* behind the data.

For example, in response to a remote enquiry, a computer may send back the message:

13.37 121314 106 104.5 106.4

This is a piece of data which might be telling us that equity share 121314, whatever that is, on the Paris exchange, at 13.37 p.m., is currently standing at 106 and has a day low and high as shown. Very useful, if you already knew the format of the reply because this automatically provides meaning and hence informs. But suppose the computer had sent something like this:

0001001100110111 0001001000010011000010100...

In fact this is the same set of digits that start off the message shown above, except they are in a different *code* (each decimal digit has been converted into a *BCD* digit, i.e. we are employing a method that uses binary as a code for decimal digits, the basis of the EBCD and EBCDIC data storage codes).

Even if you know the significance of the original six items of data, i.e. the order in which they were sent back to you, unless you know that the second example is in BCD and you can read BCD, you will not get any information.

So we could say that data is information without meaning, the meaning being supplied, either because of an explanatory set of rules, or because of the way in which the data is presented. To illustrate this latter point, a user-friendly terminal screen for the share price message would be presented so that the enquirer was in no doubt as to the meaning, perhaps something like that shown in Figure 2.1.

The dialogue mentioned earlier can be thought of from two aspects. On the one hand, it is the means whereby we 'inform' the computer (via the software) what functions it is to perform and what data it is to act on. On the other, it is the computer

```
┌─────────────────────────────────────────────────────────────┐
│                  PIERRE'S BOURST OF DATA                     │
│                                                             │
│        Date: 27th June 1986 Time: 13.37 PARIS Exchange      │
│                                                             │
│           121314 WIDGETTES s.a. Ordinary share             │
│                                                             │
│           Current buying price 106.00Fr                     │
│                                                             │
│        Low for the day: 104.50Fr High for the day: 106.40Fr │
│                                                             │
│       PLEASE ENTER NEXT SHARE NAME (RETURN FOR MAIN MENU):   │
└─────────────────────────────────────────────────────────────┘
```

Figure 2.1   *Typical screen dialogue*

(under the control of appropriate software) examining data and informing us of the results of its processing.

The word 'dialogue' is similar to the word 'conversation' in many respects and if you look at the mechanics of a conversation, there are several problems that require solution and questions that need answers. If you think about it, you will realise that the questions below actually cover almost all of the problems in communications:

● How can we send information/data over a distance?

● How do we know we are speaking to the right party and is he awake anyway?

● What if the other party has access through a system that is very different from ours?

● Who speaks first?

● Is there a requirement for simultaneous conversation?

● What if the other party is not available and the message must be sent now, or there is a local time difference?

● What language is to be used and will the system provide inter-translation if sender and receiver speak different languages?

● How can we be sure that a message has been received?

● Has the message been corrupted by the transmission procedures?

● Could it have been intercepted (bugged)?

● Could someone else have made use of my facilities?

● Is the volume of information/data large and is speed of transfer critical?

● Could the conversation have been carried more cheaply?

● Was there a better way to communicate?

● What if the sender speaks faster than the receiver?

● Can the sender talk to several receivers simultaneously?

● Can a sender select in advance the people he wants to talk to or not answer?

● How can use of the system be logged/accounted for?

● How can you control the use of the system, reduce costs and prevent staff from abusing it?

In order to select and satisfactorily implement communications systems, questions like these must be answered. Bear in mind that the protocols and standard usages that are gradually coming into general usage are designed to cope with these questions although the degree of conformity is still not that high.

For the rest of this chapter, we will be going through the wide range of concepts and techniques that are the basis of modern communications.

Types of conversation

The sender and receiver for voice represent in the first instance, a simple conversation, although when we consider conference calls, recorded messages, etc., it can quickly become much more complicated.

Incidentally, the term 'conversation' will be used even if only one party wants to 'talk', since communication in general does not have to be two-way when you are making a phone call.(You expect the occasional grunt or 'Oh, yes' from the called party even if only to acknowledge that he is still there.)

Before introducing the essential difference between voice and data transmission, we might consider a range of 'conversations' and identify the 'parties' involved without looking just yet at how the interconnection is made.

Terminal-to-terminal The name terminal is given to any unit that is at one end of a link, e.g. you sending me a telex, or our fax machines talking to each other. A more complicated example is a manager using an electronic diary to send a request for a meeting to several other managers.

Terminal-to-processor Examples are: two word-processing clerks in a local area network, both accessing a standard paragraph from a floppy disk file; a stockbroker hooking into Pierre's Bourst from his PC; a bar-code reader picking up product code/quantity sold from a supermarket checkout and sending it to update the company/branch stock file; a user at a factory site signalling the head office computer that he is ready to have sent the monthly production material requirements to his local VDU or printer.

LAN-to-LAN A company might have a series of small-area LANs within a particular site, perhaps one for the word-processing pool, one for a group of spreadsheet users and so on. It makes sense for an accountant to be able to send instructions and data into the other LAN so that a clerk can wrap up a financial investigation into a report. The company might also have a subsidary which has a very different kind of LAN, and connection between the two might be needed: perhaps the situation is 24-hour program development where a team in one country works their shift and then passes the intermediate results to a team in another country so that they can carry on.

LAN-to-WAN The LAN for the earlier word-processing pool might need to talk to a corporate videotex system, which itself has gateways to telex, viewdata, etc. A LAN for really effective electronic use will almost certainly need gateways to external services (fax, viewdata, etc).

WAN-to-WAN Multinational companies often have networks in each country in which they operate, driving smaller LANs and isolated terminals. These networks will be linked so that financial and technical data can be shared internationally. For example, there could be a central store for engineering designs from which any company in the international corporation can have a copy faxed to them or displayed on a VDU or graph plotter. Similarly Interpol might exchange fax images with the FBI.

The link between the two parties in a conversation is called a *line* and the overall connection is called *simplex* if it is all in one direction, *half-duplex* (or *switched* duplex) if conversation is possible in either direction, but not at the same time, and *duplex* or *full duplex* if simultaneous conversation is possible. The next aspect is the number of actual or conceptual 'wires' needed for the line. In general, there will be at least two wires along which information is sent and received. Additionally, there might be further wires used for sending and receiving control signals relating to timing, error-checking, protocol conversion, etc.

If a 'wire' carries more than one signal, we say that it can accomodate more than one *channel* (you will see how *multiplexing* makes this possible).

Transmission modes

As far as data is concerned, we can distinguish between *serial transmission*, where the bits of data are transferred one after another, i.e. serially, along a single channel, and *parallel transmission* where the data is sent down several channels in parallel (at the same time).

If we assume for the time being that we are sending/receiving data in groups of 8 bits, there are two fundamentally different approaches. We could send the eight bits 'abcdefgh' along one channel, in serial (bit-by-bit) mode or along eight channels, each bit at the same time, as shown in Figure 2.2.

Obviously, serial transmission is cheaper in terms of the channels needed, but is slower because the bits are sent one by one instead of all together. However, there is a problem with parallel transmission that does not arise with serial. Due to external disturbances and small imperfections in the electronics, as the transmission distance increases, there is a greater chance of the individual bits of a parallel burst of signals not reaching their destination at the same time. If severe, this 'creep' could lead to the message becoming unreadable.

Most long-distance transmission is therefore serial in nature, parallel transmission usually being restricted to very short distances such as between peripherals and a local computer.

Synchrony

Assuming serial transmission, bits will be sent one at a time from sender to receiver and, whatever happens during transport through the system, they must be synchronised or 'in step'. It will otherwise be impossible to extract meaning from the message. This must be a constant process because, however closely the two ends match, there will always be some slight differences which eventually will lead to loss of synchronisation.

With *asynchronous* transmission, the sender and receiver are resynchronised at the start of each character sent by means of 'stop/start' bits and the gap between characters can be ignored. However, the stop/start bits are bits just like those that form

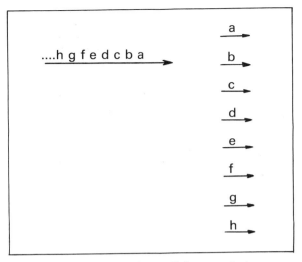

Figure 2.2 *Serial and parallel transmission*

the character and assuming that eight data bits are needed for each character, the addition of two stop/start bits means that 2/10ths of the channel capacity is wasted as far as data transmission is concerned.

It applies particularly to low-speed terminal transmission, where there are delays between the generation of characters, e.g. sending a piece of keyed-in text from a word-processing clerk to a remote disk store, or using an on-line badge-reader to send flexitime readings to a remote payroll computer.

For *synchronous* transmission, timing marks are inserted around blocks of characters instead of 'framing' each character with stop/start bits. In this way, data is sent faster because there is less overhead. The 'clock' or timing marks may be inserted by the modem or can be sent separately.

The technique lends itself to applications where characters are naturally presented as blocks, such as the output from a terminal that sends lines or pages of text, or particularly when such a terminal is receiving data files or text from a videotex network or a fax transmission.

In general asynchronous transmission is slower than synchronous and although acceptable over the PSTN, leased lines and networks will normally require synchronous transmission for speed.

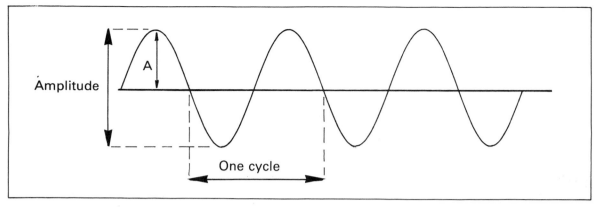

Figure 2.3 *A sine wave*

Voice and data transmission

As sources of information, sound and 'computer' data are fundamentally different. Speech is passed by a form of energy that is *continuous* in nature in that for the duration of the sound, there are no gaps. In addition, it is *variable* in the sense that it is recognised by the fact that it is continuously changing. A pure musical note is detected by the human ear as a certain number of vibrations per second, i.e. it has a certain *frequency*. The low-pitched mains hum you can sometimes hear when a radio is not properly earthed has a frequency of 50 hertz where 1 hertz (Hz) is one cycle per second. For domestic TV, test signals are generated, such as when the TV channel goes off the air at the end of the evening. Two signals are used, the higher one being about 10,000 Hz which is a fairly high-pitched whistle.

Figure 2.3 shows some of the fundamental terms used in relation to variable signals.

The diagram is an *analogue* of sound in that it is a graphical representation of the way that the energy level varies in time. Mathematically, it is said to represent a *sine wave*. The horizontal axis represents time and the variation in time is shown by the height of the curve above or below the horizontal axis. This kind of trace or *wave* is often called an analogue signal. The maximum vertical level (A) is called the *amplitude* of the wave. With a waveform like this, the element that repeats is called a *cycle* and the *frequency* of the wave (or signal) is that specified by the number of cycles in one second. The name Hertz (abbreviated to Hz)

is given to the unit 'cycles per second'. Two signals are said to be in *phase* if they have the same frequency and their peaks and troughs occur at the same times even if they have different maximum/minimum values. A phase change is said to have occurred if they get out of step.

Many types of energy are wave-like in nature and, as a form of energy, a burst of a single frequency corresponds to a single, pure musical note. Speech is a combination of sounds, originated and generated by the human brain and hopefully intelligible to another and all sound (and noise) is a combination of many single musical notes, so the actual waveform may appear to be quite random (see Figure 2.4)

A *continuous* signal is one that has no breaks in it. The source of the signal does not stop throughout the transmission period. For most of the applications this book is concerned with, data is generated and has to be transmitted in a form

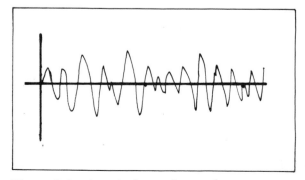

Figure 2.4 *A typical speech waveform*

that is not continuous. We say that it is *discrete* in that it is broken up into pulses rather than being a continuous wave. Also the vertical only takes on certain discrete values, usually a maximum and zero, rather than being any value between 0 and the amplitude.

If we think of 0 and 1, for example, as representing the two extremes, these are *digits* and we can say that the data is *digital* in nature. This does not mean that data is always digital. A system might have been set up to analyse the noise of low-flying aircraft and the detector/collector must be able to handle the continuous signals. In fact, it is usual to *digitise* such data anyway. (A typical technique is to take samples as digital readings at regular time intervals or an average value over a time period.) Until fairly recently, this distinction between digital data and continuous sound was important. Data is usually generated in analogue form and, for transmission purposes, the traditional serial telephone line is not very good for transmitting digital data, neither without distortion nor at a very high speed. The voice lines that make up the *Public Switched Telephone Network (PSTN)* are designed to handle a frequency range of 0.3 to about 3.3 kHz and as a continuous signal. This is usually considered as a nominal frequency *bandwidth* of 4 kHz.

When we look at the sending of data (figuratively as ones and zeros), we are expecting the line to be able to switch between two levels, one representing '1' and the other representing '0', ideally as shown in Figure 2.5. In practice, because of the electrical properties of the line, the switch from one level to another is not instantaneous (you can think of the system as having a kind of elasticity) and the signal will look more like that shown in Figure 2.6. In other words, the square edges are 'rounded off' and if this effect is too great, the detector at the other end may not be able to sort out a '1' from a '0'. It would seem obvious to abandon the idea of sending 'square' signals altogether and make use of a continuous signal, so that for example, one cycle represents a '1' and another represents a '0'. (This is the basis of *modulation* which we will look at shortly.)

For speech transmission, the brain is pretty good at sorting out the signal from background noise and even at quite low 'signal-to-noise' ratios, where whole words are lost, it is still ossible or a

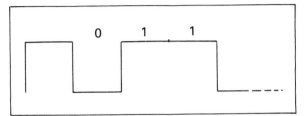

Figure 2.5 *Digital data*

Figure 2.6 *Digital data distorted during transmission*

con sation to carried out. With data, which is almost always sent in some coded representation, as zeros and ones, it only needs one bit (binary digit) to be lost and the whole message could be ruined.

There have been two major attempts to overcome this problem of error distortion. The first has been the implementation of a number of error-detection and error-handling measures which we will deal with. The second, both for this and other reasons has been the introduction of *packet switching*, both in LANs and WANs.

Instead of a *point-to-point* connection between the two parties, which is a real link between them for the duration of the call, the messages are broken up into packets of a limited size, which then go into the transmission network independently to be re-assembled before being passed to their final destination.

Line speeds, bandwidth and modulation

Going back to what was just said about telephone lines not being very good for high-speed data transmission, for reasons of cost, voice lines have a limited *frequency bandwidth* because of the electrical properties of the standard kind of line selected (largely due to costs). This is why telephone sounds very artificial – the low and high frequency components of the human voice are chopped off.

This frequency restriction (limited bandwidth) also limits the speed with which digital data can be sent down the line. It is a scientific fact that the narrower the frequency band, the slower can the change from one state to another be detected and hence, the fewer zeros and ones in a given time period. This is why that in the deeper levels of the telephone system, messages are transferred to higher frequencies (as we will see later on). The rate at which a line can switch states is called its *baud rate* and 1 baud is one state change per second.

As you can see, particularly with reference to asynchronous transmission, referring to the speed of a line in characters per second is very misleading because some of the capacity is used up in the transmission procedure itself. In addition, it is possible for more than two states to apply, which can mean that a state change could involve more than one bit (see 'dibits' shortly). A common measure of transmission speed is *bits per second* (bps) or multiples such as *kbps* (1000s of bps). Transmission rate should not be taken to be the same as baud rate which will be explained as soon as we have introduced modulation. But you must appreciate the difference between 'bits per second' and 'baud' because the two are used rather sloppily in practice.

To introduce modulation, let us look at the output from a keyboard after a user has entered, say, the character 1, which might be in an ASCII 7 bit group. The bit pattern for the character '1' looks like this:

0110001

We build our input devices that generate this pattern of electrical pulses perhaps as different voltage or current levels to distinguish between 0 and 1. So to send the character 1, we want the receiver to pick up something (from left to right) as shown in Figure 2.7.

For the telephone transmission of sound, we speak into the hand-set and the microphone converts sound energy into an electrical analogue , i.e. an electrical waveform that mirrors the variation in time of the sound. This is sent by 'attaching' it to a fixed frequency *carrier wave* (see Figure 2.8).

This attachment is carried out electronically by a process called *modulation*. The idea is for a 0 or 1 from the data signal to affect the carrier differently and there are several ways to do this. The first is called *amplitude* modulation (AM), where having synchronised the zeros and ones with cycles of the carrier, a 0 has no effect on the carrier, while a 1 might halve or double the *amplitude* of the carrier for the cycle in which it coincides. The effect is that a 0 has a certain loudness and a 1 will have half or twice the loudness. The detector must be able to pick out this difference in loudness in order to recreate the original signal pattern. Now you may be able to understand why the frequency band limits the baud rate. Staying with loudness for a moment, we can think of the overall range of loudness as the equivalent of the frequency band, the narrower the difference between the two levels to represent

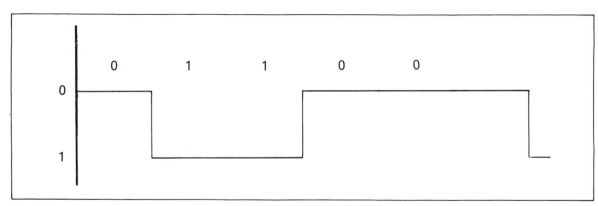

Figure 2.7 *Part of the character '1' in ASCII*

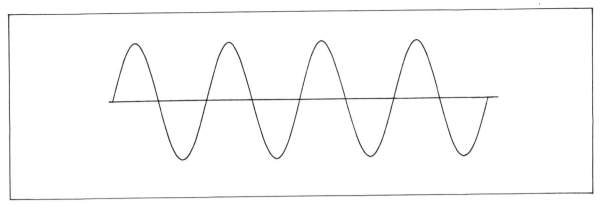

Figure 2.8 *The 'carrier' wave*

the 0 and 1, the harder it will be to separate them as the transmission rate increases. The modulation is as shown in Figures 2.9 and 2.10. The modulation technique has halved the amplitude of a cycle for a 1.

A *more common* method of modulation is where one of the digits changes the *frequency* of the carrier wave for a cycle. This is, of course, called *frequency modulation* (FM). In practical terms, the 1 and 0 are represented by different musical

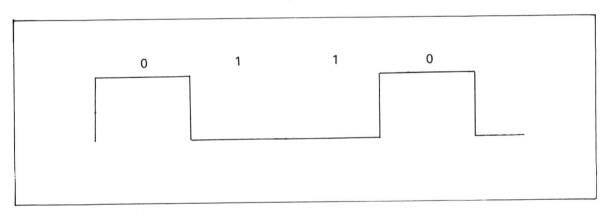

Figure 2.9 *Original bit pattern*

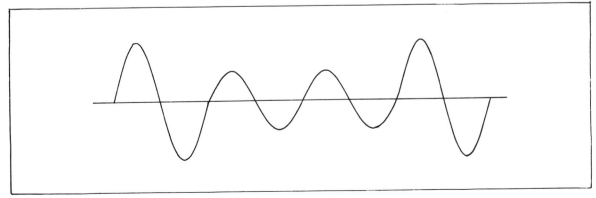

Figure 2.10 *Carrier amplitude-modulated by the signal*

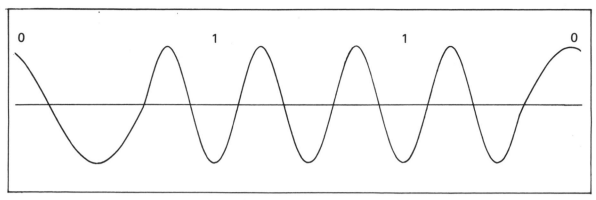

Figure 2.11 *Carrier frequency-modulated by the signal*

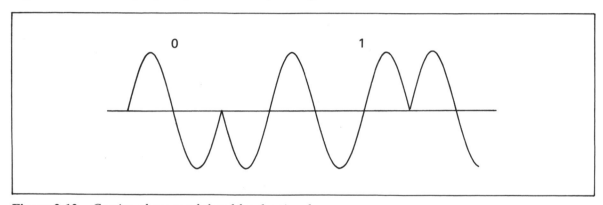

Figure 2.12 *Carrier phase-modulated by the signal*

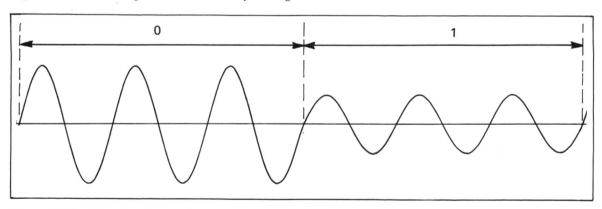

Figure 2.13 *Amplitude modulation*

tones. The result of modulation of the above bit stream with the same carrier wave might be something like that shown in Figure 2.11. As you can see, the 1 has doubled the frequency of the carrier (there are twice as many cycles in unit time). The effect of this modulation is to produce two different tones, in effect, two musical notes, rather than two levels of loudness.

To see the effect of frequency band on transmission speed, try whistling a digital pattern to somebody, first with the notes an octave apart, then with just a semitone separation. The receiver will find it very difficult to pick up the semitone difference especially at speed. Whatever modulation technique is employed, the separation of the two states (loud/soft, high note/low note, etc.) must be increased as the transmission speed is to be increased. The problem is even more obvious if you use two levels of loudness in the whistling.

The third common modulation technique, called *phase modulation (PM)* sets cycles out of phase to represent 0 and 1 (see Figure 2.12).

Other modulation techniques, more sophisticated than AM, FM or PM are available. The names, for the record are *pulse code modulation (PCM)*, *pulse amplitude modulation (PAM)* and the more recent *adaptive differential PCM (ADPCM)* and *continuously variable slope delta modulation (CVSDM)*.

You should appreciate that the views of modulation given are somewhat simplified, in that the bits of the signal actually modulate more than one cycle of the carrier. If we consider a device generating 1000 bits per second, for a 3000 Hz carrier, there will be three cycles to carry each bit. For example, amplitude modulation in this situation would look more like that shown in Figure 2.13.

Some of these more sophisticated modulation techniques can actually reduce the number of bits that need to be transmitted. One method looks for changes from 1 to 0 and vice versa rather than actual 1/0 values.

PCM is becoming more important in transmission because it makes it possible to send continuous data in a kind of digital form. This seems paradoxical in that the earlier methods are all concerned with the problem of converting digital data into continuous.

Essentially, an input analogue signal, which can be speech or data or both, is sampled at very short

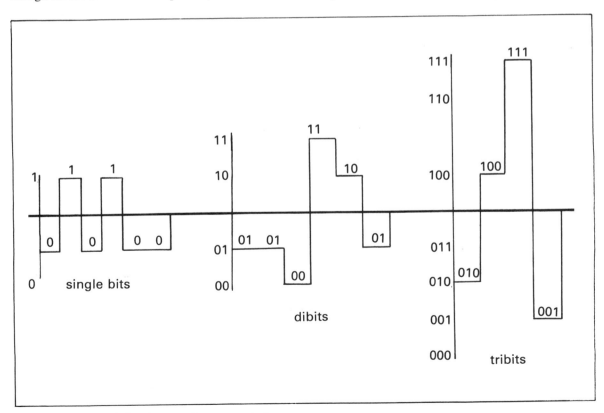

Figure 2.14 *Multilevel states for transmission*

intervals, the interval usually depending on the frequency bandwidth of the line (as we have seen, typically 4000 Hz) and since the sampling rate must be at least at twice the highest frequency, the sampling is carried out at 8 Hz (8000 times a second, or with an interval of 125 microseconds).

Each sample corresponds to a pulse which is then passed to a coding unit which examines the height (size) of the pulse and after comparison with 256 standard heights, converts it into a binary code, i.e. a number consisting of binary zeros and ones. Each one of these can then be sent as a pulse down a data line or converted into two audio tones corresponding to the zeros and ones. In fact they are more likely to be converted into pairs of tones as you can hear if you have a touch-phone linked to a digital PABX. As you press a key, you will hear a combination of two tones. The handset contains a device called a *CODEC (Coder-decoder)* which converts into tones and back again. Because the actual pulses are very short in comparison with the sampling frequency, they can be time-division multiplexed by slotting them onto a common channel.

So far, we have referred to just two possible states having the significance 0/1. Using the various modulation techniques it is possible to have 4, 8, 16 or more states in what is called *quadrature amplitude modulation (QAM)*. This could be obtained with four or more levels of loudness for AM and four or more tones for FM. This means that we can send a modulated signal where the data transfer capacity is higher than the baud rate.

For example, with four states, we could assign significances 00, 01, 10 and 11. To send the message 010100111001 with two states would need 12 state changes, whereas with four states, we can send 01 01 00 11 10 01, which only needs six state changes. In other words, for a given carrier frequency (and baud rate), we can send twice as many *dibits* as we can send bits. Grouping into sets of *tribits* would treble the transmission rate since we would send 010 100 111 001 corresponding to four state changes (as shown in Figure 2.14).

Baseband and broadband

These are two terms that appear both in LAN and WAN technology.

Baseband implies that the message signal is sent so that transmission is within the frequency band of the transmission source. Over short distances (perhaps up to a few kilometres), the various distortions and losses that make long-distance baseband transmission difficult are much less important and, provided we accept that, for reasons mentioned only slow transmission rates are possible, such as for voice, teletype and micro-to-micro communication.

Broadband implies a channel that has a frequency range wider than is needed for the sending of one signal. Thus, within the overall band, sections of the frequency range can be allocated to several different messages at the same time; in other words, making use of frequency division multiplexing. For example, in the PSTN, 12 4-kHz voice baseband channels are carried on a broadband from 60 to 108 kHz, so that one is from 60–64 kHz, the next from 64–68 kHz, and so on. Even higher up the network, the 60–108 kHz can be thought of as another baseband and this and other 60–108 kHz basebands could be included in an even larger broadband which could itself be a baseband for yet an even higher broadband.

To illustrate this, think of a group of messages, A,B,C,..., which are carried on a 60–108 kHz band, as shown in Figure 2.15. A different set of baseband channels could be multiplexed onto Y, another 60–108 kHz band. Then X, Y, etc., could be be multiplexed onto an even higher frequency band and so on. Hence, as the frequency band increases, the transmission capacity becomes higher. This is exactly what does happen in practice, with the increasingly high transmission rates being carried along optical fibre or microwave links (dealt with later in this chapter).

Matching unequal speeds

There will be many occasions when it is important to find a way to equalise devices that 'run' at different speeds. For example, a company might have a central computer, on-line to its branches, each of which has data entry and enquiry terminals which are used intermittently. Typically, a telephone line from each individual terminal to the host could be as slow as 110 baud. It seems crazy to have each terminal on its own 110 baud line and we can avoid this by using a *multiplexer (mux)*,

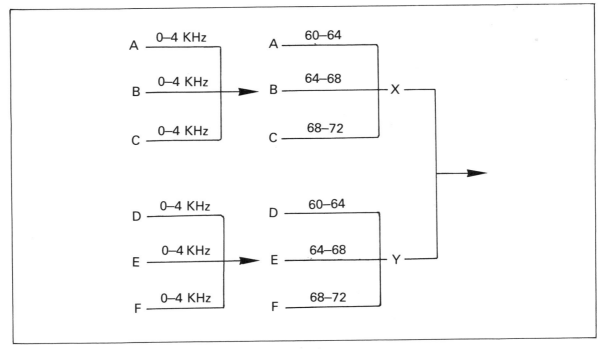

Figure 2.15 *Multiplexing onto a broadband channel*

which we can use to connect six terminals via one line rated at 300 baud. The mux basically scans the attached terminals and picks up or delivers a character at a time to/from each one in turn. Since they are in use intermittently, they are unlikely all to want service at exactly the same time, so a baud rate of 6×110 is not necessary. Multiplexers are used for this purpose in LANs, private and public networks.

A more complex beast is the *Statmux* or *statistical multiplexer*. This is rather more intelligent than a basic mux in that it tries to improve the general level of multiplexing. For example, during the 'polling' or scanning process, when it is sending or receiving characters, it can keep a record of any terminal which has been idle for a while and perhaps miss it out for a few cycles. This means that the time to poll the terminal is not wasted and can be better employed with the terminals that do need service.

Obviously, at the other end, their must be a *demultiplexer* synchronised to the original mux to separate out the components of the combined signal. There are two basic multiplexing methods, the first uses time and is called *TDM (time division*

multiplexing). In this, the mux takes a piece of data from each terminal in turn so that for example, for *n* cycles, it is sending data from A, then for the next *n*, from B and so on.

A different approach can be employed where several slow- to medium-speed channels need to be linked to a very much higher speed channel. The technique is called *Concentration* and, in its simplest form, this just means bulking up slow data until there is enough to send at a faster rate in bursts. The bulking can be on magnetic disk or in ROM. With the increasing use of microprocessor technology, concentrators are becoming more sophisticated and intelligent.

Modems and other data transmission hardware

This area contains enough subject matter for several textbooks, but we can at least give some idea of the main items and their function.

Modems

Having already mentioned modulation, let us start

with the *modem*. This essentially takes digital data and impresses it on a continuous carrier so that it can be sent in the analogue form required by the telephone system. A similar modem at the other end will extract the original signal from the modulated carrier. The name for the device comes from *MOdulator–DEModulator.* The simplest form of modem is connected by 'dial-up', i.e. the terminal equipment is linked to the telephone line via the modem and to establish contact with a distant receiver (computer or other terminal), it is necessary to dial his/her telephone number. The receiving equipment will then reply with a signal showing that it is ready to receive data and the sender manually switches to 'send data' mode. All very nice, but of limited application. Many users will be making use of a direct line leased from the common carrier and what we need to make use of it is an *autodial/autoanswer* modem which can connect out as soon as it is switched on and can be activated by an incoming call. The line leased can be of normal 'voice quality', i.e. they are the same lines as used for voice from telephone to exchange in the PSTN. The main advantage is in the convenience of not having to dial out. These lines are limited to about 2400 baud.

For use with dial-up services, modems are usually asynchronous. With the higher-grade lines, synchronous modems are rather more efficient and generally have the ability to attach timing marks to the data. Asynchronous transmission is handled by software. Higher-grade 'special voice grade' lines can be leased which are of use up to 9600 baud and 'wideband' lines are available for higher-speed transmission. In a later chapter we will look at the packet switched system and digital transmission facilities, which allow for much more efficient data communications than is possible over voice-grade lines.

An example of an autodial application might be a data collection terminal such as a cash point, which may need to transmit a statement request or a balance enquiry at any time of the day, but which would be expensive if it was connected continuously. Similarly, autoanswer would be needed in electronic mail or telex if automatic handling of incoming messages was wanted.

An alternative for modem use is the *acoustic coupler* which uses sound for modulation. Once connection has been made to the opposite end, the

Figure 2.16 *Modular Technology Minimodem 3005 accoustic coupler*

Figure 2.17 *BT Merlin DM4962X modem*

terminal is connected to the coupler which generates different audio tones corresponding to the zeros and ones of data from the terminal. These are then passed into the telephone which sends them just like voice to the other end. The telephone handset is pushed into rubber mountings on the top of the coupler. Owing to the fact that they are rather cheaper than other modems and because of possible background noise on the line, the error rate is higher than a modem, but their advantage is that they are portable and can be taken with a terminal and used with any telephone.

BT as the main supplier of the PSTN facilities have been supplying modems for over 20 years, but since liberalisation, many companies can offer modems in a vast range of types and speeds.

Modems are rated by the speed with which they can handle data (usually as bps – bits per second) and can be conveniently classified into up to 300,

300–1200, 1200–2400, 2400–4800, 4800–9600 and 9600+, e.g. 14,200, 19,200 and upwards.

BT have recently launched their *Merlin DM4962X* modem, which has packet-switching facilities and can handle duplex operation at up to 9600 bps over a two-wire circuit.

Note by the way that in half-duplex transmission, the commonest speeds are 300 baud in each direction (300/300), 1200 baud in each (1200/1200) and 1200/75 where the two channels have different speeds, the faster one for the message and the slower being used for message confirmation, or, more likely, the slower for requests and the faster for data transfer.

The largest market at present seems to be in the 9600–19,200 range. Sales of lower-speed modems are starting to fall off because of the reduction in price of the high-speed ranges. There seem to be about 250 companies offering modems in the UK, including IAL, Data Communications, Dynatech Communications, Thorn-EMI, Case, Hasler, Racal, Perkin-Elmer and, of course, British Telecom.

When looking for a modem to support data communications it is important to know something about the communications software that will be used and the standards to which it conforms. When installing a large data processing, videotex or other data transmission network, the communications software will be provided by the network supplier or with his cooperation. For the small user, care must be taken. As we have seen, there is an increasing trend to use a PC as an intelligent terminal for database access, telex, electronic mail and for integrated packages where for example, a user might want to send a spreadsheet screen to another user. Although these facilities can all be accessed through the same modem, it is necessary to use different interfacing software for each.

For example, integrated packages like Lotus, Symphony and Framework have inbuilt communications software which can be tailored to suit quite a large range of different modems. Plessey and Racal-Milgo supply modems that fit inside an IBM PC and extra software facilities come with packages like *Crosstalk*, *Datatalk* and *Comm* which allow access to integrated packages, word

processors, Telex, viewdata, etc., by menus that you can set up yourself.

We will say more about this when we look at protocols and standards but, in particular, note that you should know a little about some of the CCITT recommendations in the range V21–V29, which apply to modems up to 2400 bps.

Smart modems

Some modems, particularly those developed by the US company, Hayes Microcomputer Products Inc., and the standards that they have set for their industry, are called 'smart' because they can do 'lots of clever things'. As part of their ability to autodial/autoanswer, they can respond to a series of commands, for which they can be 'trained', i.e. programmed. Unlike other types of modem, where options are selected with 'dip' switches, the Hayes *AT* autodials commands set them by software. They are called the 'AT' set because each command is preceded by the ASCII characters 'A' and 'T'. They can store a message in a buffer and analyse it. From this they can detect whether transmission is synchronous or asynchronous, spot the transmission speed and, in a wideband situation, if a particular band is noisy, they can switch to another. They can also handle loop disconnect or touch tone exchange signalling (dealt with in a later chapter) and deal with a busy/disconnected situation or where a voice answers, and password/answerback, where a password is received and the caller rung back if it is acceptable.

Backus Systems in the US, offer a product called *Dialcontender* as an extension to the standard modem for computer access at a host computer. When a user dials in, a standard procedure accepts an identity and user number and calls back when it has verified the information. It then links the user into the host. If there is no available connection, the user is given the option to join a queue or be rung back when one becomes free. This is not a standard modem activity but shows just what is being provided these days to automate data communication within a Hayes command-like framework.

The Hayes *Smartmodem 1200*, which was the one on which the AT set was originally implemented, is a particular autocall/answer product available in the UK, together with *Smartcom II* software for PC users. It is claimed to be fully compatible with CCITT V22 for full duplex, asynchronous communications and can be switched in and out to/from voice without breaking the connection, so that a telephone discussion could occur and then data transmission continued. It also operates in half-duplex. As with certain other intelligent modems, this product incorporates test facilities that enable it to look for and detect errors in the linkage. The machine has a display panel so that call progress can be monitored and a command set that enables it to be used in different situations, such as with different PABXs.

The Miracle Technology *WS3000* and the Steebek *Quattro* were among the first to be given BABT approval as multi-speed modems. The Quattro 2400 is a Hayes-compatible modem and handles 2400, 1200, 1200/75 and 300 bps on one card for the IBM PC. Crosstalk mentioned above is tailored for use with it.

Problems with the use of modems

There may be big problem areas for modem users. For example, in asynchronous mode incompatability at each end can be caused by the minutest difference between timing signals generated and interpreted. Some modems have facilities for fine-tuning (the Hayes options mentioned are really a form of coarse-tuning).

Another problem that may become more common as the number of suppliers increases is that they may claim that their communications software is compatible with a particular microcomputer or host system. However, many machines have their logic on a PROM card and as requirements change or manufacturing processes develop, the card may be changed. If you have an earlier version, you may be in trouble. This could be the case with a 'Hayes-compatible' modem and 'Hayes-compatible' software where the modem can handle extensions to the original 1200 command set but the software has not been tailored to cope with them. The obvious answer is to insist on a demonstration before purchase.

In the next chapter, you will find reference to a British Telecom service called *Multistream*, which is relevant when thinking about terminal and modem selection because it offers an advanced degree of error-handling.

Other devices

Generally, modems are for *long-haul* use, i.e. for longish distances (more than a few hundred metres), since a modem is a digital-to-analogue device and the amplification needed will only operate on analogue signals (digital amplifiers are much more difficult to use and are more expensive). Other devices are employed over short distances. They include short-haul or baseband modems, modem-eliminators, amplifiers and line drivers or repeaters. Short-haul modems need a wide baseband but are useful over short distances because there is no need for amplification.

With a standard micro-to-peripheral link (such as an RS232 cable between PC and printer carrying out what is essentially baseband transmission), there is a limitation of about 50 feet on the connection distance. This is not usually a problem in an office, but you may want to share facilities over a factory site, for example, where one person makes use of an integrated package and wants to share files with another PC on the other side of the site. One possibility is to fit a modem to each and just make use of a telephone line as the link. In fact, provided all users have the same communications software and very similar modems, there is no reason why a local area network could not be set up on this basis, except for the cost of the modems and the use of the line.

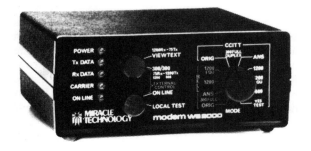

Figure 2.18 *Miracle Technology WS2000 modem*

Figure 2.19 *Modular Technology Microdriver line driver*

Provided you are using a package which has some communications software, a cheaper alternative might be to link without a modem and there are several possibilities. An *amplifier* can be used to increase the power of a signal where there have been losses due to the electrical properties of the transmission medium (called *attenuation*). A *line driver* can also be used to increase signal strength, but acts rather differently from an amplifier in that it *regenerates* the signal.

If a weakened, distorted signal reaches an amplifier, it will increase the signal strength and boost the distortion at the same time. A line driver will analyse the incoming signal, determine which are ones and which are zeros in the general mess of signal and noise and then produce a stronger set of corresponding signals, at the same time eliminating the noise.

The line driver is not always the complete answer because its use introduces a delay in the system and, in addition, it only works in one direction, so you need one at the other end. It is, however, valuable when distances are fairly short and you don't want to use a modem.

Note that the BABT impose severe restrictions on where a line driver can be employed outside the local exchange area.

Easydata Communications Ltd offer a range of line drivers which handle from one to seven channels and are recommended for use within a building, where it is important for VDUs and printers to be sited quite a long way from the host processor. A very recent product they call *Dataexpress*, can handle transmission rates up to 200,000 bps.

Modular Technology Ltd offer a range of modems and line drivers, in particular a very neat line driver called a *Microdriver* which is only 85 × 45 mm in size and which can handle 19,200 bps for up to 1250 metres, or greater distances at lower speeds.

A *modem eliminator* is particularly used for testing equipment and has facilities for providing variable timing signals and can often generate or simulate error conditions. It is also used when synchronous devices which need an external clock are connected locally. In other words, a pair of modem eliminators at each end of a link can simulate connection across a telephone line.

Multiplexers

Multiplexing can apply to voice and/or data messages, but is probably easier to explain in terms of data only. Essentially a multiplexer allows multichannel operation over a signal transmission path and an early application of multiplexing was in mainframe computers where, in addition to fast magnetic disks and tapes, a number of slow-speed devices were used (card readers, paper tape readers, printers, character readers, etc.) and it was not economical to give each device its own individual link to the CPU. The fast devices would be connected by their own fast 'selector channel' (IBM terminology), while all the slow ones could share a 'multiplex channel'.

By 'byte interleaving', the CPU could send or receive individual bytes to/from each device in turn using the technique of *polling*, in which on a regular, 'democratic' basis, each device is interrogated to see if it is ready to send/receive the next byte and a byte at a time the data is received/sent.

Figure 2.20 shows two input devices and one output device connected to a CPU through a multiplexer – the bold characters are output from the CPU, others are being read in. This is, in fact, an example of time-division multiplexing and, as we have seen, it lends itself to situations where a number of very slow inputs, e.g. word processing

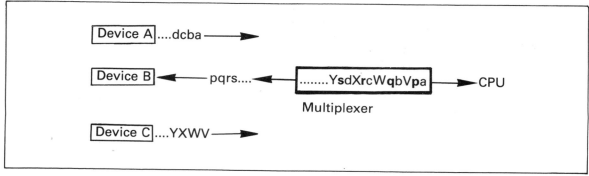

Figure 2.20 *An example of multiplexing*

or electronic mail typists at about six characters per second or say 100 bps, can more economically share a line at 300 bps rather than have their own line.

Some manufacturers offer multiplexers that carry out TDM by interleaving *bits* rather than bytes. Easydata Ltd sell what they call their *Datatruck* range, the fastest of which can handle up to 134,400 bps along screened cable up to a distance of 5000 ft.

Frequency division multiplexing is the obvious way to make effective use of a wide bandwidth and transmit data in parallel, i.e. instead of chopping up messages into bytes which are interleaved, different frequency bands carry several messages simultaneously.

British Telecom, amongst other companies, offer their *Netmux Series III* based on TDM in connection with their high-speed *Megastream* service (see next chapter). The machine can handle up to 30 different speech and data time-slots at a variety of transmission rates on the 2 Mbps line.

From a very recently published buyer's guide it would appear that the majority of multiplexers offered in the UK are based on FDM rather than TDM.

Statistical multiplexers

Quite a number of companies sell statistical multiplexers based on STDM (Statistical TDM). The advantages to be gained from the statmux stem from its intelligence, i.e. its computing power based on an associated microprocessor.

Essentially, the statmux is built to handle 'real' signals only, in that it is clever enough to use all the available slots to carry active messages. The mux detects that a source is idle and does not send null data. In more fancy terms, the *statmux* deals with 'demand-based allocation of channels using statistical techniques'.

There are however, three other factors that lead to increased efficiency. The first is the ability of the mux to compress data before it is multiplexed and transmitted, in particular by converting the more common characters into codes shorter than the eight bits required for ASCII or EBDCIC. It does mean that some of the less common characters will take more than eight bits, but on average, one code in particular will give an overall reduction in the number of bits that need to be transmitted for a particular message by over 25%.

The second aspect of an intelligent statmux is the provision of some memory, usually called a *buffer*, which is used to store and hold part of the data in an *overbooking* situation. If more traffic passes in than can currently be handled, some can be stored until the mux is able to analyse it. The effect is that the mux can in theory allow a gross input greater than the capacity of the line that it is connected to, perhaps up to a 6:1 ratio or even higher.

There is some overhead associated with the use of the mux and the way that it operates. It groups data into 'frames' which need identification and routing data and some extra 'redundant' data attached for error-detection purposes (explained later in this chapter). The efficiency of the various framing techniques used depend on the volume of traffic to be handled, but all are aimed at sending only active data.

The third aspect is the way that the mux can handle the various protocols associated with data. Most statmuxes handle asynchronous data, although some can support synchronous protocols. A particular frame-assembly method strips off the timing marks which reduce the overhead of asynchronous data. Some statmuxes are capable of monitoring and reporting on excessive error rates and other 'system statistics' such as the efficiency of line and buffer usage with a time attached to any message generated.

The overall result of employing a statmux is that an uneven traffic flow in time does not mean that the equipment must be geared to the maximum and, at the same time, the statistics gathered can show over- or under-use of facilities.

One final aspect is the way that they can be linked together so that data is 'switched' between them. If a particular mux is malfunctioning, another one may bypass it and send the messages along an alternative route.

Because of their extreme sophistication, it is possible to select a varying number of channels, mixing data generated from widely differing sources, perhaps handling, at the same time, LANs, personal computers, videotex and telex.

Data transmission media

Twisted wire

This has been the commonest form of transmission medium over very short distances, such as in local area networks and particularly from telephone handset to PABX. It usually consists of a pair of sheathed wires twisted together within an outer sheath. Resulting from the possibility of different lengths of the two wires, there may be slightly different times for transmission which can give problems. Furthermore, although transmission rates can be very high (Mbps), twisted pair is very susceptible to electrical disturbance, but it is not

difficult to use since for very sensitive areas, such as near heavy machines, it is possible to break the pair and join in a piece of coaxial cable or optical fibre, which are shielded from interference.

Coaxial cable

Coaxial cable (or 'coax'), which is principally the same as is used for TV aerials, is essentially a line for transmission consisting of an inner conducting wire surrounded by a dielectric (non-conductor) usually made of plastic, although some large cable could be air- or gas-filled. These are surrounded by an outer sheath making up another conductor. The basic coax cable is a special development of the twisted wire pair and can generally support much higher transmission rates (up to 10 Mbps) handling perhaps 100 voice channels. Coax is used in local area networks (to be discussed in Chapter 7) and is used in telephone networks mainly because the noise level is somewhat lower than for high-frequency radio links, even though more expensive. Coax will also be used in areas where a high level of high-frequency and microwave transmission is used, since this would otherwise cause severe problems.

It is also used when there is a very heavy use of the channels, leading to multiple coax, which can be cheaper than many radio links. Yet another use is in connection with TV or video-conferencing and with the more recent developments in technology, it is becoming possible to multiplex voice and low-definition TV. A major difficulty with coax (as with fibre-optic cable) that is eliminated with radio and microwave, is the physical problem of laying cables, not forgetting the right-of-way that must be obtained before cable-laying can take place. There is also the possibility of frost damage if the cable is laid too near the surface and, of course, accidental damage (where someone puts a shovel through the cable.)

Repeaters are needed at successive intervals, which can be buried with the cable or may be on the surface. To make life easier, cables are cut to length at the factory so that they span from repeater to repeater.

For duplex transmission within a wide network, a pair of cables is needed, although there is a standard single cable which can accomodate 120 voice channels in both directions. A single coax is called a 'thin' cable and where the outer sheathing encloses several coax cores (2, 4, 6, 8 or more), it is called a 'thick' cable. Where more than one core is involved, spacers are inserted to separate them.

With frequency bandwidths of 50 MHz plus, it is quite possible to carry many thousands of channels, voice and data.

Microwave

For radio broadcasting, various frequencies are used depending on the quality of reception required. Signals are modulated onto a carrier frequency in a manner somewhat similar to that used in modems. FM provides better quality transmission than AM, but with data, where the error level is very much more critical than with speech, radio broadcasting is both inefficient and expensive.

In addition, as we have seen, to carry enormous numbers of channels, we need very wide frequency bands and even when dealing with the VHF (very high frequency) FM radio band (80–110 MHz), there may not be sufficient capacity.

There has been an increasing use made of *microwave*, which is a form of radio at a much higher frequency still (1000s of MHz). Unlike lower frequency transmission, microwave is not broadcast but is directed from point-to-point (transmit to receiver) along a path through the air called the 'line-of-sight'.

Engineers would say that broadcast signals are 'unbounded' while directed signals are 'bounded'. A consideration here is that broadcast signals can be picked up, in theory, by anybody who has detection hardware, while if directed, there will not be much chance of 'tapping in' except by picking up any side radiation. With lower-frequency microwave, it is possible to pipe the modulated signal to the transmitting aerial along coaxial cable, but for higher transmission rates (and a correspondingly higher frequency), it is necesary to send it through a metal pipe with a rectangular cross-section called a *waveguide*. The

dimensions of the rectangular inner are very carefully engineered to be an exact number of multiples of the wavelength of the carrier (centimetres).

Over very short distances, it is posible to send the microwave along a waveguide, but as soon as this becomes more than a few metres, the cost of laying an accurately made waveguide becomes prohibitive and is obviously out of the question when we want to send a batch of telephone calls, say, across a city. There are two methods of concentrating the microwave signal before transmission. The first and more common is to make use of a 'dish' aerial (or antenna). This is usually a saucer-shaped 'dish' to which the signals are conveyed by coax or waveguide, and which by reflection can send a parallel beam to the receiving site. The curvature of the dish is specially designed to achieve this concentration effect and the diameter, like the waveguide, is a function of the wavelength. With microwave and its extremely short wavelength, dishes can be of the order of metres in diameter. With lower frequencies, the dish would have to be enormous.

For high-power work, the waveguide itself can theoretically be extended to form the antenna on the rooftop. It is drawn out to form a *horn* shape which by its design concentrates the signal, thereby eliminating the need for a dish. (The horn is not used much now.)

Being of such a high frequency, line-of-sight microwave is not affected by electrical disturbance from machinery, nor from electrical storms.

While on the subject, it is worth referring to the similar dishes used to link with satellites. The nature of communications satellites is outside the scope of this book, but we can anticipate the section on Mercury Communications in Chapter 3 who have a satellite link in London. Figure 2.21 shows the dish used to transmit their 'Americall' service to the USA. In the background are the microwave antennae that 'backhaul' messages to the national network with line-of-sight connection to the City of London.

Mercury have also installed a larger (18-metre) dish, supplied by Marconi, at Oxford. It provides the basis for their leased-line service to Hong-Kong. They expect to install a 13-metre dish to

Figure 2.21   *Mercury antenna, 13-metre, at East End Wharf*

replace the 18-metre dish which will be used for services to the USA.

Optical fibre

Light, as with most other forms of energy, is wave-like in nature and can be used as an incredibly high-frequency carrier for data and voice. Modulation is carried out electronically using PCM and the resulting pulses are applied to a light emitter (either a laser or a light-emitting diode) which produces flashes of light corresponding to the electronic pulses applied. These are then passed down the optical fibre and the detector converts them back into electrical signals.

The basic fibre is a very fine hollow 'wire', the core being made of silica (which is chemically similar to glass) or certain forms of plastic. This is surrounded by a 'cladding' of a different material and because of the different optical properties of the core and the cladding (the core has a lower refractive index), a light beam is continually reflected from the sides of the core until it emerges at the other end, with no loss by absorption of the signal into the cladding. There will be losses every time a reflection occurs and any impurities in the cladding could also lead to some losses, but there should be absolutely no side radiation from the fibre.

The physics of fibre optics is way beyond this book but we can just say that related to the wavelength of the light, the internal diameter of the core and the refractive indexes of core and cladding, light signals are dispersed which can severely limit the transmission capacity of the fibre.

Monomode fibre has an extremely fine diameter which makes it very efficient as a transmision medium but until fairly recently, has been difficult to join. The alternative was to make use of *multimode* fibre by juggling with the refractive indexes of the cladding so that it is not uniform along its length.

Another problem is the fact that the light sources used are not a single frequency, but a small range of frequencies. Even a laser generates a small range of frequencies unless an expensive one is used. This will also limit transmission capacity because the different frequencies are reflected to a slightly different extent.

Since we are dealing with reflection, transfer can only be in one direction. Signals from both ends would interfere, so for duplex use, a pair of fibres is required. Even this problem is being solved by utilising the technique of *wavelength division multiplexing* which allows full-duplex along a monomode fibre. These days, BT and Mercury make use of multiple-fibre 'bundles' and seem to

Figure 2.22 *Jointing optical fibre*

have overcome the jointing problem. Figure 2.22 shows jointing under microscope, of monomode fibre of 125 millionths of an inch diameter, used in Mercury's network in London.

Most fibres used in the main Mercury networks have 10 fibres, although the 'London Ring' which follows the British Transport Underground Circle Line is based on more.

Mercury fibre-pairs are rated at 140 Mbps, but they claim that up to 560 Mbps is possible, allowing for 8000 voice channels. One of the big advantages of optical fibre cabling is its small diameter and weight, which means that it can be laid fairly conveniently under floorboards without the need for the 'false floor' required by heavy cabling. This has been put to good use by several companies. The London merchant bankers, Lehman Brothers, have a large complex of 'desks' with different kinds of dealers accessing terminals and the facility to go through the whole building with fibre-optic cable supplied by Pilkington.

The massive use of fibre optics enables Charing Cross Hospital to link up with six other large London hospitals with Mercury microwave point-to-point dishes.

Another advantage of fibres is that signals are not affected by high voltage from machinery.

A very recent development by BT has been the application of 'blown' techniques in the laying of fibre cable. The empty plastic outer sheathing is laid along the chosen route using convenient factory-cut lengths which can be connected by means of special coupling pieces. Then using a special device which carries the fibre by means of two rubber wheels, it is introduced into the opening of the sheathing. Once there, compressed air is used to send it along, inside the sheath. If a resiting/relaying operation is necessary, the fibre can be 'blown out', the sheathing can be re-laid and the fibre blown back into the new installation.

Very recently BT announced that they are starting work on a 'global optical-fibre network'. It intends to set up a complex of undersea optical-fibre cables to form a global network of high-capacity digital links from the UK to Europe, the USA and the Far East. The systems will start with a 120-km link, called *UK-Belgium 5*, from Kent to Belgium carrying almost 12000 channels over three pairs of cables each operating at 280 Mbps.

TAT 8 scheduled for summer 1988 will be based on similar cables and will link Cornwall with the USA and Brittany with the USA. *UK-Denmark 4* is also scheduled for 1988 and BT has co-ownership of *Transpacific Cable 3* between the USA and Japan (also for 1988) with a Hong Kong extension for 1990.

BT have recently ordered over £1.7m of fibre-optic data communications equipment from Focom Systems as part of their Customer Services Systems network. It will be used to provide data links from remote Megastream multiplexers to IBM 3274 network controllers. Until Mercury and BT started using fibre optics heavily, its use was largely restricted to local area networks and we will return to this subject in Chapter 7.

Laser transmission

This is of limited application, but is worth mentioning for short-distance work, both for data, voice and video. Modular Technology distribute their *Interlaser* which is recommended for distances up to 1 km. Signals are modulated onto an infra-red beam (similar to the light used in fibre optics, but with a lower frequency). The transmitter looks very much like a video camera and is fitted with a range of attachments which provide for adjustments and fine-tuning. (The company also sells a similar device for which the light source is a light-emitting diode. This is designed for up to 200 metres.)

Error detection and correction

There are many techniques employed in communications and data processing for detection, correction and prevention of errors, equipment failure and mistakes. In data transmission, the term *redundancy* is often used. This means basically, sending extra data above that actually required, which can be used for error detection. The techniques are not really relevant to voice and for bulk data, such as file-transfer, or the remote transfer of collected transactions to a mainframe, the occasional error can be handled in a fairly simple way in that a character with a parity failure could be converted to a character that should never normally occur. The data once received could then be scanned for this character and the relevant records retransmitted at a later stage.

But with database access or, say, in a reservation or bookings situation, errors must be detected and some form of request-repeat applied immediately. The same applies when 'downloading' software, i.e. transferring machine code programs from one machine to another. With data, an error is often obvious, but when sending true binary data as is the case with machine code (or perhaps large binary or floating-point numbers), almost any combination of bits is possible and an error might not be detected for a long time.

Echo checking

This is the simplest, but is not very economical. The receiver sends back the message in chunks as it is received and the sending equipment then compares what was sent with what was echoed back. If there is a difference, the original is re-sent. The technique is generally of more use when dealing with data transfer between mainframe and peripherals.

Parity checking

This is often called *redundancy checking* and is the best example to explain of the various techniques. If we consider the sending of a group of eight bits (usually called a *byte*) down a line, if one of the bits is set the other way, perhaps due to an external disturbance or hardware failure, the receiver will get a distorted message with no way of knowing that an error has occurred. Consider the example shown in Figure 2.23. Note the strange way the bits are numbered: there is a good reason for this. In computing, it is usual to number from 0 upwards, and the right-to-left is because the bits in a binary number represent powers of 2, increasing from right to left. If for example, the 4 bit is set to a 1, the receiver would get distortion as shown in Figure 2.24. Similarly, setting off the 0 bit would generate the distortion shown in Figure 2.25.

To detect this kind of error, we attach an extra bit (redundant in that it is not actually needed to identify the character) that is set as a function of the number of ones in the original character.

Taking the EBCDIC 'D' (11000100) and *even parity*, the extra bit is set to a 1, so that the total number of bits is an *even* number (110001001). If any bit is set the wrong way during transmission, the parity will no longer be maintained and the

Bit number 7	6	5	4	3	2	1	0
Bit setting 1	1	0	0	0	1	0	1

Figure 2.23 *The character 'E' in the code EBCDIC*

Bit number 7	6	5	4	3	2	1	0
Bit setting 1	1	0	1	0	1	0	1

Figure 2.24 *The character 'E' distorted to give 'N'*

Bit number 7	6	5	4	3	2	1	0
Bit setting 1	1	0	0	0	1	0	0

Figure 2.25 *The character 'E' distorted to give 'D'*

Character	ASCII	Sent as (even parity)
@	100 0000	100 0000 1
A	000 0001	000 0001 1
M	100 1101	100 1101 0
h	110 1000	110 1000 1

Figure 2.26 *ASCII characters with parity bit attached*

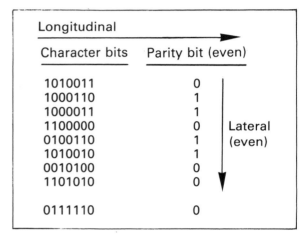

Figure 2.27 *Parity checking in two directions*

electronics in the receiving terminal will detect a *parity failure* and request a retransmission. Similarly, with *odd parity*, the number of ones should be an odd number.

The ASCII character code has provision for parity checking in that it is based on a 7-bit structure. Data is usually sent in bytes so the extra bit is set for parity checking (see Figure 2.26). The disadvantage of this basic form of parity checking is that it will not detect an even number of errors – if two bits are set the other way, the parity will be unchanged. This kind of parity checking is *longitudinal* in that it acts along the bits that make up the message. We can also apply *lateral* checking, so that, for example when sending eight characters, we could make up a parity bit for all the corresponding bits (see Figure 2.27).

If any one bit is set on or off, its corresponding longitidinal or lateral parity bit can be used to pick it up. If two failures occur in one line, although its parity bit will not be affected, two others will, so any combination of errors can be detected.

Checksum (cyclic redundancy checking)

This is a more complex technique in which a mathematical operation is carried out on groups of bits of the message sent. In this technique we treat each bit or byte as a digit and to divide each by a number (obtained from a 'generating polynomial' and find the sum of the remainders. At selected intervals, this sum is sent to the receiver that is performing the same calculation. A discrepancy indicates a failure.

ARQ (Automatic Repeat Request)

These redundancy-checking techniques are employed in the more general procedure called *ARQ*, in which the system either *stops and waits* for a reply ('ACK' usually indicating a successful receival and 'NAK' meaning that a repeat is required), or is *continuous* in that the sender continuously sends and receives signals. When a 'NAK' is returned, the sender starts transmitting again. This may be *pullback* in which the sending system backtracks to the part that failed and retransmits from there, or *selective-repeat* in which it only re-sends the part in error. Generally, stop-and-wait ARQ is employed in half-duplex links, while continuous ARQ applies to full-duplex.

Forward error correction

This is another possibility which can be applied in several different ways and is much more complicated than the techniques already mentioned. What usually happens is that the original message bits are converted into a different code which is designed in such a way that the receiving circuitry can not only detect errors but has the ability for *Maximum Liklihood Decoding*, i.e. it can 'guess' what the original data was with a greater degree of probability, depending on the construction of the code.

Chapter Three

Networks and exchanges

British Telecom, Mercury and liberalisation

First of all, the larger part of this chapter, and the book as a whole, is concerned with a description of many facilities and services offered by British Telecom. Until *liberalisation*, British Telecom (or the Post Office as it was then part of) had a monopoly position as the only provider of telephone network services and no company was allowed to compete apart from Hull Telephone Department (part of Hull City Council) which was given a licence to operate by the Post Office in 1904.

Mercury Communications Ltd was originally set up in 1981 as a consortium of Cable & Wireless, BP and Barclays Merchant Bank but since August 1984 is now wholly owned by C&W.

Mercury is able to offer switched and leased line services in competition with BT and can also interface with BT networks. Its main difference from BT is that it is mainly concerned with providing transmission paths while BT is also into equipment such as modems, exchanges, etc.

So, although BT's name appears frequently in the text, it must not be forgotten that they are no longer the only public telephone authority in the UK.

Before looking at telephone systems we should look at a quick summary of the legal environment in which telecommunications operate. After many years of total monopoly in the provision of telephone and telecommunications facilities, in 1981, the British Telecom Act was enabled. This separated the old Post Office into the Post Office,

which would continue to handle the postal services, and BT which would be only concerned with telecommunications. In addition, companies were now allowed to compete in the telecommunications hardware and services market. This overall relaxing of monopolies and the encouragement of fair competition has been called *liberalisation*.

In a later Telecommunications Act (1984), Hull City Council, Mercury and BT were given licences to act as *Public Telephone Operators (PTOs)*, as which they would be able to offer basic services over fixed data transmission links. These are the only companies that can offer public switched networks directly to the public or to companies, subject to the conditions under which their licences were offered.

The relationship between Hull and BT is friendly because Hull only operates within a clearly defined area and presumably will never compete with BT on any other basis. Whereas there was obviously initial hassle with Mercury because BT felt that their delicate pricing structure and competitive edge could easily be smashed by a competitor entering the field and offering unrestricted 'bucket-shop' services. However, this was satisfactorily resolved in 1985 and Mercury is now perfectly entitled to supply a public network which can link into the BT network and can offer switched services internationally (it has been very successful in taking advantage of this in introducing powerful services to the USA and Hong Kong).

Administration and supervision is in the hands of the Department of Trade and Industry. The

Telecommunications Division is responsible for the licensing of the various networks offered whatever transmission medium they are based on.

(The Radio Regulatory Division deals with radio services, such as the allocation of broadcast frequencies which used to be the responsibility of the Post Office).

The 1984 act had several far-reaching effects (in addition to setting up the initial PTO licences) since it enabled the privatisation of BT at the same time of course, removing from it, its power to issue licences to others.

It was also realised that cable networks, such as had already been offered by 'cable-TV' were before long likely to become important in general data transmission and it granted licences to several cable operators. It also granted licences to the cellular radio operators, Vodafone and Cellnet. Equally important was the establishment of *The Office of Telecommunications (OFTEL)* as an independent body which would oversee the fair operation of licence holders to ensure that the telecommunications industry provides decent facilities and which assists industry and the economy in general. (An OFTEL '*Network Code of Practice*' should be available soon.)

To regularise the sale and maintenance of services and equipment, the *British Approvals Board for Telecommunications (BABT)* was set up. It is quite independent of BT and the other PTOs and its responsibility is to ensure that equipment such as switches, modems and telephone handsets are of a standard that allows them to be connected to and used in services provided by the PTOs. Approval is shown with a circular, green label and a product that is banned from use has a red triangular label.

Since 1982, the *Value-Added Network (VAN)* and its licensing conditions have been defined. This is essentially a network service which offers more than a means of transferring information/data between users – such as providing some storage facility, such as electronic mail, or altering the content or form of messages over and above that needed for transmission by the network. VAN operators require a licence to enable them to offer their services based on transmission facilities provided by the PTOs.

To start with, very few companies applied for licences, but about 150 VAN operators now offer their services. We will say more about VANs in the appendixes at the end of this book.

The telephone system

A conversation, whether voice or data and whether one or both ends are active, involves messages going through some network and there must be some means whereby the system can recognise that the caller wants service and then provide a route from the caller to the called party.

The network that handles all the calls is involved with several quite different functions:

Signalling   The means whereby information can flow so that the network knows who needs connection to whom, for call-charging, monitoring and eventual call close-down. To link normal analogue exchange lines to the public network a number of basic trunk signalling systems have been used (AC13, AC15, DC5, DC10) and the more recent MF5. The latter has not become a standard, but with the expanding demand for connection between networks, the increasingly complex exchanges in use and the extra facilities wanted from telephones and telephone services, signalling now needs to send very much more information than was formerly provided.

With systems where several call channels are multiplexed onto one line, it is usual now to employ *common-channel* signalling, where one channel carries the signalling information for all the other channels.

An example of signalling is the ability of the system to be able to turn on/off the charges for a call (the BT *Linkline* service enables calls to be made free or at local rates depending on the dialling code).

A recent development has been the introduction and adoption, as interim standards in this country, of the common-channel signalling technique for private networks, called *Digital Private Network Signalling System (DPNSS)* and a similar standard for public systems *Data Access Signalling System (DASS–2)*. These signalling methods provide a basis on which individual PABXs can exchange signals at high speed relating to call set-up, holding and clear-down as well as facilities like conferencing. It is supported by BT and Mercury as well as most of the large suppliers of switches and it is being used in several large private ISDN. Plessey claim that a DPNSS version of their IDX switch (discussed later on in the chapter) was launched as early as last September and about 28 systems delivered in the following month.

The main effect of DPNSS is that it makes possible 'network-wide transparency' of the features available from the PABX, in particular the possibility for most of the 'intelligent telephone' facilities like diversion, 'call-back-when free', etc. (These facilities are dealt with in detail in Chapter 4). It also provides the ability for PABXs of different types to communicate. In particular, 32–channel lines are used.

Switching   The mechanism that actually provides the connection between caller and called party via intermediate nodes.

Routing   Deciding on the actual transmission path for calls.

Transmission   The processes whereby voice and data are safely and cost-effectively transferred in a suitably coded form to their destinations.

Flow control   Ensuring that the load on the network is spread out evenly and that error conditions are handled so that the system is run efficiently and, hopefully, cost-effectively.

Switching and transmission

There are three quite distinct mechanisms for switching calls. The first, which has been in use for many years is called *circuit switching* and means that a physical link is set up along an electric circuit, perhaps consisting of different media, and this link is continuously held until the call is terminated. The second is *message switching* where messages are entered and held on what has always been referred to as a 'store and forward' basis, until routing has been established and the called party is ready for the message. The third is called *packet switching* and differs markedly from message-switching in that, as you will see later,

there is no need for a fixed connection and this leads to a very efficient service.

The telephone lines, connection circuitry and the devices that route calls and link callers and called, make up the *public switched telephone network (PSTN)* which is currrently based largely on analogue transmission.

The main lines along which calls are routed all over the country are called *trunk* lines and the subidiary lines which connect the trunk lines to various exchanges are called *branch* lines. Voice calls cannot be treated in the same way as data calls for the very simple reason that a voice call must be continuous. The receiver is essentially the human brain which is picking up the message serially, i.e. word by word, at a certain rate. With data, for example a screenful of information from a database enquiry, the individual characters could be sent any old way, perhaps in several separate chunks with a time delay between them. Provided the formatting software can re-assemble the message and lay it out on the VDU, the receiver can read it at his leisure.

For voice and low-volume data traffic, the analogue-based service is effective even though a cerain amount of distortion and disturbance occurs. With voice the sound is very restricted, because of the 4 kHz frequency band and background noise is very common. However, the human brain is fairly good at picking out sense from even a high level of surrounding noise (low signal-to-noise ratio) so we usually get by. The situation is rather different with data, in that just one bit lost, i.e. turned on/off or flipped the other way, could ruin a whole document transmitted. Fortunately, as we have seen, there are several techniques that can be used to spot and handle corruption in data.

Transmission problems

Noise on lines is due to several causes and we can identify at least four of these:

- Increasingly more voice and data traffic is being carried so there is more to disturb and more to be disturbed.

- There are several different types of exchange

in use: the Strowger, Crossbar and SPC we will meet shortly. As a result there is some degree of interference/incompatibility between them.

- The increasing use of electrical devices, such as motors and switches, outside the network and the proliferation of other transmission networks leads to local interference and disturbances.

- Many different electronic devices are used within the network such as amplifiers, line drivers, multiplexers, etc., and these all introduce various distortions and losses due to their incompatibilities and differing electrical properties.

One point that shoud be realised is that BT obviously can make no guarantees about line integrity for *data* on dial-up lines which were designed for voice transmission. This means that they and you must recognise the impossibility of eliminating errors on existing voice lines (although some of the error-handling 'redundancy' techniques discussed in the previous chapter will reduce them). If you want a much lower error level, you will need to lease a line or a whole network from BT (or Mercury). What characterises low-volume traffic is that, by definition, it works in short bursts, separated by gaps. During a phone call, there will be pauses for thinking before a reply. During data entry, the user will probably spend time looking at a source document or data validation error messages relayed from the host computer receiving the data. Of course, the disadvantage with a permanent link during the call is that the connection must be paid for even while thinking is in progress and techniques such as multiplexing, concentration, etc., have been introduced to try and make line usage as efficient as possible. Otherwise, BT could not possibly handle the traffic – while a circuit is switched to a particular call, it is obviously not available to other callers. Furthermore, the data error-detection techniques are effective, but all require more bits to be sent in one way or another.

The operative word is 'intermittent' and since the conversation is intermittent, the line spends a large part of its time doing not very much. If there is noise, it occurs when no-one is listening.

But once we start dealing with high-volume traffic,

e.g. passing files between computers, using fax to transmit complex drawings, or sending complex videotex screens from host to enquiry terminal, the intermittent nature of the transmission largely disappears. Since noise can interfere at any time, the effects are more severe, especially since we are also dealing with faster transmission rate.

If we are sending at up to 110 bps for 10% of a typical session (the other 90% being for thinking time), a burst of noise taking even a second or so has a very good chance of destroying none, or perhaps only one character (8 bits!). But if we are sending at 9600 bps for 70% of the time, a noise burst of even a tenth of a second could wreck a hundred characters.

What is really needed is a method of sending data efficiently in terms of minimising the data overhead or (redundancy) needed to detect errors, but also minimising the possibility of errors disturbing the data. At the same time, it would be nice if we could make use of the various techniques like multiplexing, for carrying many messages along one path (or channels along one line), thus eliminating much of the transmission medium and at the same time getting rid of all that electrical circuitry needed in circuit-switching. These aims are achieved in the services provided by System X and packet switching, dealt with later on.

For voice transmission, the means of connecting or routing calls was through a series of central junction points, called *exchanges*, in the trunks, with operators at these switching centres and manual connections. (These days, the switching exchanges, whether in a public network or in one set up for use within a private organisation, are called 'switches'.)

Then Strowger exchanges were introduced to do this electromechanically (and very clumsily by modern standards although for its time, it was very ingenious). The next improvement was the Crossbar exchange. This exchange is electrical/electromechanical in nature, but still rather slow. The problem is that dialling codes have to be translated from the general STD code into a series of local codes in order to activate local switching centres.

Dialling There are two different methods of sending dialling information to the exchange. The older method, geared to the telephone dial, is called *loop disconnect* and each digit dialled causes a corresponding number of make/break signals (clicks) to go down the line. The more modern technique is called *Multifrequency (MF)* or *Dial-Tone Multifrequency (DTMF)* and is geared to push-button dialling where each digit generates a pair of tones that are close together, the two frequencies representing each digit being selected for minimum error by the signalling circuitry. In either case, the signalling pulses are passed into the system which uses them to establish connection with the party dialled.

Although you are using a push-button handset, it does not necessarily mean that DTMF is implied. The push-button is for convenience and unless you are connected to a private digital exchange or a digital network, you are still 'dialling', in that each button causes a set of clicks to be generated. In fact, with a push-button telephone connected to the PSTN, it seems that establishing a call takes longer than before – perhaps 10 seconds. This is an illusion really. With a dial, the *post-dialling delay* takes about 1 second per digit to dial and then the connect time. The push buttons can be pressed in, say, 2 seconds, but it still takes the same amount of time to connect the call.

If in use with a digital exchange, the telephone will actually generate tones to the central exchange which will then signal the network in the usual manner. Obviously, when a public digital network is available, MF signalling will be direct from phone to network. Similar times are needed to close-down and disconnect a call once finished. There is some overhead in ensuring that the connection is maintained during call progress.

The post-dialling delay may be critical. With voice, a few seconds waiting time is not really a problem, but if you are cycling data to many different terminals, the call set-up time to establish each of the connections may be several seconds – much greater than that required for data transmission. With the new digital systems, connection times may be reduced to as low as a quarter of a second.

Physical connections The physical connection from telephone to exchange through the socket is based on a two-wire 'pair' which can support the typical typical duplex human conversation.

Higher up the network, this two-wire working becomes impractical. The first problem is the electronic losses that result from the physical characteristics of the line. Amplification is thus needed to bring the signals back up to a recognisable level. The amplifiers used can only work in a one-direction circuit (two inputs and two outputs) so at some stage the two-wire coupling will need to be converted into four-wire. Equally important is the fact that the 0.3–3.3 kHz baseband (nominally 4 kHz) for speech is insufficient for more than one message unless time-divison multiplexing is used, i.e. sending the bits of individual messages interleaved in time. This is not practical for speech because of the delay it would impose. In addition, TDM is not efficient for analogue transmission. It is much more effective in local area networks based on coax cable or fibre optics.

In the existing, non-digital PSTN, a typical line will have 12 channels imposed on it, each with its own baseband of 4 kHz (slightly more than needed for the 0.3–3.3 kHz) using frequency division multiplexing based on a wide band of 60–108 kHz, so that one message is carried on 60–64 kHz, the next on 64–68 kHz and so on.

If coax is employed, an even wider range of 1 MHz can be used, thus carrying in theory, 1000000/4000 or 250 channels. Even higher up the network, the channels can be sent down fibre-optic links after conversion from electronic pulses into extremely short flashes of light, or via a microwave dish. In both cases, TDM is used.

The basic exchange topology was revised around 1970 to take advantage of more efficient signalling and switching techniques, but has undergone major revision since the introduction of fibre optics and microwave and, later on, with the extensive usage of computer control via System X.

The trunk transit network

This is the name given to the overall system whereby exchanges at different levels are connected together. The first point of access is a *local exchange* from which a call can be connected along the same single-pair to another user within the local exchange area. At a slightly higher level are *group switching centres*, which convert from two-to four-wire signalling. The group centres are partially interconnected while a local exchange will only link to one or two group centres. A higher level still includes about 27 *district switching centres* which are strongly interconnected and which go into 9 fully interconnected *main switching centres*.

Switch technologies

If we now go inside a building or some other site containing telephones it is obviously expensive for each telephone to have a separate line and exchange unless there is a definite need and the *private branch exchange (PBX)* was developed to handle the switching of calls internally and to hook any particular extension to the outside. In its original form, the PBX was manually operated and was called a *private manual branch exchange (PMBX)*.

The earliest types of switching equipment (about 1920) were electromechanical and are still in use to some extent. The first is the *Strowger*, (*TXS* or *step-by-step* switch) with which each digit dialled causes a mechanical selection, the final route being selected by the combination of digit selections. The other switch (from about 1930) in common use is the *Crossbar* or *Common control* switch in which the switching elements are built up into a matrix of horizontal and vertical bars. The digits making up the number dialled are received by the switch which then activates the switching elements at the same time, rather than step by step.

The big problems with electromechanical switching units, apart from their lack of sophistication, are associated with difficult maintenance and the fact that moving electrical contacts get worn, dirty and work loose. This latter can cause considerable disturbance and is probably the major factor contributing to line noise and interference.

It became obvious that some electronics technology was badly needed and spurred on by wartime research and the explosion in electronics that came with the transistor, switches started to become more cleverly constructed and more powerful in operation. They were able to take advantage of several techniques like multiplexing, which are really impossible with mechanical switching.

These days, the unit is automatic and in fact can be a very powerful piece of kit, containing the same kind of processors and other electronics used in microcomputers. In fact, they may be referred to as having *stored program control (SPC)* and have now been around in the UK for over 10 years.

A more recent factor has been the development of digital services where the modulated analogue signal (voice as well as data) is converted by PAM into ones and zeros which can then be sent as tones, just like DTMF dialling. Digital technology is employed both in local switches, i.e. in a building, and in whole geographical networks.

The general trend is shown in Figure 3.1, which shows the development of switch technology from all electromechanical through electromechanical telephone with digital switching to fully digital switches. It is expected that even more sophisticated digital switches will soon appear which will interface with digital trunks and will behave like small networks themselves.

Trends

Some figures may help to demonstrate trends in telecommunications. Experts claim that voice traffic will increase at the rate of about 12% per year and data traffic at almost twice this rate. (Annual inland call growth for BT has averaged 5% since 1981 and international growth about 12%.)

Electromechanical	Electromechanical + electronic	Digital
Strowger (TXS) Crossbar (TXK)	SPC (TXE)	System X System Y etc.

Figure 3.1 *The development of switch technology*

The number of electronic exchange switches in use by BT rose by 50% from 1982 to 1985, but the number of digital switches rose from 1 in 1982 to over 80 in the same period.

In the main BT trunk system, 77% of the links are coaxial cable, 14.5% microwave and the rest optical fibre.

The PABX

The name given to the local unit, to which extensions are attached, is *private automatic branch exchange (PABX)* and it is worth saying something about the PABX because it is of value not only for telephone call switching, but also for data, both in wide and local area networking. It should be pointed out that the term 'switch' can be confusing since it implies something fairly small. In fact, the very largest of telephone exchanges from Northern Telecom, Thorn-Ericsson, Plessey, etc., are all called switches and the term PABX really now only applies to quite small set-ups.

A typical PABX set-up consists of the wiring, various telephone extensions, various frames (points of interconnection) and the PABX itself.

The main function of the PABX is to be able to switch internal calls between extensions and external calls to and from the extensions. A user gets an outside line by dialling 9. The complexity of the PABX and the quality and layout of the wiring will obviously determine the facilities available since there will need to be increasingly more intelligence in the PABX if it is to handle a complex set of operations. Similarly more wiring will be needed for them.

But the data transmission capacity and the quality of transmission will also be affected by the type and size of the wiring and how much there is of it because certain losses will occur simply due to the various electrical properties of the wiring.

The set-up is structured around separate units, called *frames* for expandability and flexibility, to allow access for testing and repair and to provide interfaces between different types of cable.

The external cables to/from the telephone or data network come in at the *building frame* and the internal cables from this lead to a *main distribution frame (mdf)* from which there will be further links or *distribution points (dp)* capable of handling 40+ pairs of the wires needed to the telephone extensions.

The next stage will be *distribution points* from which smaller numbers of pairs may emanate, eventually leading to a handset or terminal extension (see Figure 3.2).

Voice and data can be combined over the same two-wire pair, in what is called *Data over Voice*, provided care is taken to ensure that the voice does not interfere with the data. Thus, telephones and computer terminals sited near to each other could be linked within a building. The two are combined by modulating the data at a much higher frequency than the voice and then using FDM to put the two signals together.

Eventually the line goes to the distribution frame, with some kit up to a distance of 4 km along the same pair. At the distribution frame, the signals have to be separated, with voice going outside and data going to the computer. This opens up possibilities for using the exchange as a message/data switch in the building and a big saving in the cost of cabling providing the quality of cable selected is sufficient to handle both. Consideration must also be given to the cost of the separation at the distribution point.

In effect, with the variety of different services that must be provided, the PABX must really work as if it was a branch exchange within the integrated network itself and the term *Integrated Services PBX (ISPBX)* is already being used. It will also be essential for the digital PBX to be able to interface with all the specialised terminals (fax, telex, slow-scan TV, etc.), even though there is an increasing trend to multi-functional micro-computers as terminals.

A very sophisticated PABX will contain facilities for linking extensions, routing and re-routing external calls, call logging/accounting, back-up data such as list of users and facilities available and linkages to different external services such as leased lines and the various digital facilities available from BT and Mercury. In addition it can be used to handle LANs within the building.

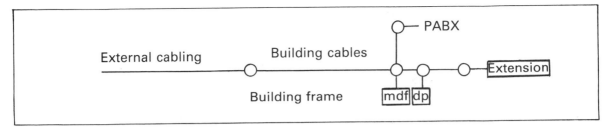

Figure 3.2 *Standard PABX installation layout*

Mercury are largely supplied by Northern Telecom who, like Mercury, are the subject of one of the appendices of the book. Northern's largest switch is the DMS250. Figure 3.3 shows an engineer carrying out pre-commissioning tests on a DMS250 installed at Mercury's new Willesden exchange.

Large switches supplied to BT are from Plessey (the DX), the XS2000 from Mitel Corporation which is now 51% controlled by BT anyway, from GEC (under licence from Northern Telecom) and Thorn-Ericsson.

At the top end of Thorn's range is *AXE10* which is the massive public digital exchange supplied to BT as *System Y*. The company expects to deliver exchanges that will contain up to 3 – 500,000 lines

Figure 3.3 *Northern Telecom DMS250*

in 1987. They say that they have already installed world-wide, equipment that deals with 4,000,000 lines with 6,000,000 on order.

There was quite a lot of trouble over System Y and after strong objections from the other suppliers, OFTEL have decided that for the next three years or so, Thorn's total contribution to BTs requirements must not go over 20%.

Thorn-Ericsson's AXE10 is very interesting and is already in operation at the heart of 'the world's largest digital international exchange', located in London. It is also used extensively by Racal Vodofone in its cellular telephone network and they have ordered two further systems.

The application requires the exchange to cope with the problem of the phone user being mobile, thereby having to decide on the nearest radio base station and exchange (more in Chapter 4). Figure 3.5 shows the modular construction of the exchange which makes testing and servicing convenient.

Cable & Wireless have recently signed a $5.6m order for the supply of an AXE for Hong Kong's new international switching centre. It will be the island's third and largest international exchange, the first two also being Thorn-Ericsson products.

Figure 3.5   *Access to the AXE10*

On a slightly smaller scale is the *MD110* digital PABX. The machine can handle voice and data/text, i.e. data processing and or videotex/electronic mail, and will support up to 12,500 extensions. The console for the exchange has five visual displays and Thorn-Ericsson claim the operation can be mastered 'in a matter of hours'. The overall system consists of a number of 'Line Interface Modules', each of which handles up to 200 extensions, and is housed in a single cabinet.

A very much smaller 'Compact MD110' can cope with up to 180 extensions. It will handle paging, abbreviated dialling, conferencing and other 'convenience' facilities. Because of its small size – less than a cubic metre – the whole exchange could be moved around if necessary in the event of a fire or office reorganisation. In fact, there is no reason why it could not eventually be sold 'on wheels'. Thorn claim that the integral battery back-up can support 7 hours of power cut. Both the hardware and software to support the system are tailored to the user. Given the type and number of operators, extensions, exchanges and private lines, a program is used to produce detailed specifications for the production and testing of the kit. It also produces an after-sales service record for the customer.

Figure 3.4   *Thorn-Ericsson AXE10 exchange*

An interesting application of the MD110 is with

Figure 3.6 *BT Monarch IT440*

the British Airport Authorities. Their aim is to set up what they call an 'Airports National Network (ANN)' and they have installed a system based at Heathrow with a main site and four satellite sites. A total of 3300 extensions will be reached with a fifth remote site at Stanstead. The main exchange hooks in 63 lines from the BT trunk and 144 direct-dial lines between BAA and clients. With an 'alternative routing' facility the system will shunt a direct-dial through the PSTN if no private line is available.

An add-on feature called 'Critical Lines Exchange' is installed to which high-priority services, such as those dealing with security and emergencies, can be switched as a back-up to the main system to ensure 24-hour a day service.

Thorn say that their system was selected in preference to a set of smaller, linked PABXs in order to have the advantages of an integrated, distributed (widespread) network such as conference calling and automatic callback.

Another major installation of MD110 is with London Transport, which links over 9000 London Underground station extensions.

Going back to the comment above about the exchange on wheels, BT have recently announced their *Commsure* caravan which is quite literally a

512-extension, mobile unit which can be delivered to a customer's premises within 24 hours of order.

It can be configured to a particular requirement and the 'caravan' contains a generator, if needed, and work space for two operators and a supervisor. The facility can be provided from within a customer's premises if necessary (no parking space?).

Commsure is offered as an emergency service to existing customers, particularly important these days where because of their small size and almost totally electronic nature, fire or water can completely destroy a whole exchange.

BT claim that, by keeping up-to-date records of the requirements of their Commsure customers, as equipment and facilities are upgraded, they can supply a mobile switch to replace exactly the system that is out of action.

BT supply a wide range of systems, some of which are 'keyphones' (see Chapter 4). Others include *Viceroy* and *Kinsman* for small systems and *Regent* and their most recent *Monarch IT440* for small-to-medium applications although the new Monarch can handle up to 440 extensions. Voice, data and text switching are offered in the same compact unit.

The TX14 featurephone can be used with all three switches.

Plessey have a very sophisticated exchange they call *IDX (Integrated Digital Exchange)* which has a 60% share of its market (private sector, 100+ lines). It is sold by BT as the *DX* and Telephone Rentals as the *TDX*.

A more recent one is called *ISDX (Integrated Switched Digital Exchange)* which supports their ISDN-access terminals and makes use of 2-Mbps circuits from BT or Mercury. It is mainly aimed at private network users right down to 10-line applications and can link to public systems where necessary. The system comes with the *ISDT 300+ (Integrated Services Digital Terminal)*, a featurephone which has ports enabling it to be connected to a computer as well as the switchboard.

The IDX, SX2000 and switches like them have

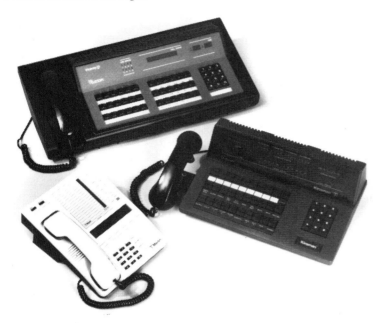

Figure 3.7 *BT Viceroy,
Kinsman and
TX14 featurephone*

Figure 3.8 *A Plessey ISDX installed in an office*

DPNSS capability and can be networked together in extremely large complexes. BT plan to introduce DPNSS eventually on the Monarch series.

The sophisticated facilities available that make life easier for the operator and the user can be provided from the internal telephone exchange while some can be obtained fron the telephone itself. We will look at them in the next chapter when we deal with 'intelligent' telephones.

Packet switching

We must eventually deal with digital transmission and its eventual logical extension into the *Integrated Switched Digital Network*, which relies on *Integrated Digital Exchanges* and for which the user will require *Integrated Digital Access*. As a lead-in to this, we must say something about packet-switching and some of the facilities available from System X. Later, we will go over what Mercury have to offer.

Unless we are dealing with the most simple of

Figure 3.9 *The Plessey ISDT digital services telephone/entry terminal*

communication links, there must be a mechanism for routing and connecting data and voice calls. In effect, the message to be sent is stored until the receiver acknowledges that he is ready to receive, whereupon the message is forwarded. This is called *message switching*.

Communication lines and circuits are available for a variety of purposes and with a variety of capacities (speeds) including the telegraph-like Datel 100 service from BT, for very low speed transmission as with teleprinters, Telex, the PSTN itself and a range of rented (leased) facilities.

BT will only guarantee data transmission rates of up to 2400 bps through the PSTN. Higher rates were possible using specially engineered lines and modems such as the older DATEL series of services, but BT must have realised that the services and transmission rates offered were hopelessly inadequate for the needs of the fast-growing 'communications industry'. Since techniques like pulse code modulation have been made available, both public and private network suppliers have been looking into the production of switching, transmission and networking using *digital* rather than analogue techniques as pieces in the jigsaw puzzle that will eventually finish up as the public ISDN.

BT hope to get rid of all the old electromechanical Strowger and Crossbar exchanges and implement (by the year 2014) a completely digital system, which will mean that all transmission is under computer control resulting in very much faster service of much improved quality. It also means that the old copper telephone cables must eventually be replaced by more modern transmission media, mainly fibre optics. It will be interesting to see just when Mercury will be able to say that they are waiting for BT to implement their own total ISDN.

An advantage of digital transmission is that signals weakened because of electrical losses along the lines can be boosted with line drivers, rather than amplifiers which bring up the noise level as well as boosting the original signal. These pick up a signal and reproduce the ones and zeros received in a way that does not pick up distortion.

Even more important is the fact that digital data transmission does not require the use of modems, since actual musical tones are generated which can be sent in this form.

BT and the companies associated with them in the development of the new telephone system realised that this could not happen overnight and after putting in the System X computers (developed by GEC and Plessey) to control the networks, they have been phasing in a digital network using coaxial cable, microwave and fibre-optic links so that now there are well over 50,000 km of 2-Mbit optical fibre which can handle a mixture of voice and data. Most of it is in trunk lines, but more is being placed in junctions.

Later on we will discuss some digital services already available and see that digital PABXs are already in use. But it must be realised that a digital PABX cannot be used to its full advantage if the transmission is over analogue lines; fully digital transmission will not be available for quite a few years yet.

However, the various companies mentioned supply fully digital switches and other equipment which are very usable in private networks both for voice and data and as a switch for videotex, LANs, etc.

Currently, a PABX based on analogue transmission can be considered for up to about 100 extensions but for much more, a digital exchange is really needed. At present, many BT exchanges are still analogue in nature. By the same token, once full digitisation has been achieved, all PABXs will need to be digital.

BT are obviously making great efforts to extend the system as far as possible and, as an interim measure pending full digitisation, they introduced a system that runs under System X called *Packet Switchstream (PSS)*, more recently called just *Switchstream*. This requires a high level of analogue transmission, but the main trunk routes handle data by digital means and they hope to convert local connections to the main trunks before the end of the century.

The technique of packet switching was not invented by BT; it has been around for some time and has become the basis of many networks both wide and local but it is so important that it is worth spending some time on the technique.

Perhaps its biggest advantage over the other means of connection is that no physical, 'point-to-point' link is needed between sender and receiver. Another advantage from the user's point of view is that several terminals can all link in to the system from the same site with less equipment than would be the case with connection to a message-switching system.

Yet another advantage is the cost of transmission. The link from the sender terminal to its nearest PAD is at normal local or STD rates, but once within the PSS, the distance travelled does not affect the charge; it is only the volume that counts (plus, of course, the charge from the nearest PAD to the receiver). The simplest view to take is that voice and data messages from many different terminals (in fact over 90% of the business telephone 'community' can access the system), are chopped up (disassembled) by a *Packet Assembler/Disassembler (PAD)* into fairly short 'packets' which are then thrown into a general melting pot (the *packet switched network* or *PSS*) under the control of the *Packet Switching Exchange (PSE)*. Each packet is created with a source and destination on it.

The fancy computers that make up BT's System X, which control Switchstream then take the melting pot, route all the packets to the PADs nearest their destinations, which recreate the original messages, passing them along the final local link to their destinations. This means that there is usually a choice of transmission paths in the event of failure or exceptionally busy conditions on a particular line. In addition, data transmission is very much faster than the PSTN.

Bear in mind that voice and data cannot be treated in the same way since the delay in receiving the re-assembled packets of the original voice call must be sufficiently low to prevent the message losing its intelligibility.

Generally, data transmission by its very nature occurs in bursts; a typical VDU/keyboard terminal may be actually transmitting for less than 10% of the time it is connected to the other end. For this, it would be very inefficient to set up the link, send the burst and then break the link. With packet-switching, the system can buffer (store temporarily) these bursts as 'packets' and send them farther down the line to another 'packet switch' containing another buffer until eventually they reach their destinations. This, of course, means that to avoid delay in transmission, packets of data must be fairly short.

The packet switch is actually a powerful computer and its intelligence is used to make decisions regarding requests to transmit, routing or re-routing, recording statistics, etc. There is always a delay in establishing the link between sender and receiver. With voice calls, we can stand this delay, but it may be very inconvenient when we are dealing with data transmission if data volumes are high or if near-immediate response is needed. Packet-switching reduces this problem. If the volume of data to be transmitted is continuous, it may be worth considering a *packet terminal*. This is a terminal that can send and receive packets directly to/from the PSS through the PSEs without the need to go through a PAD. Such a terminal is linked to the packet-switched exchange by what is called a *dataline* which can be 2400, 9600 or 48000 baud lines. The services that provide these are known by the speed, such as Dataline 1200.

Slower datalines are possible at 300 and 1200 for character terminals. It is also possible to get into the system via the PSTN using an acoustic coupler or modem, usually at 300/300, 1200/75 or 1200/1200. In this case, the character messages must be 'packetised' by the nearest PAD.

Datalines are also provided to link character terminals to their PAD.

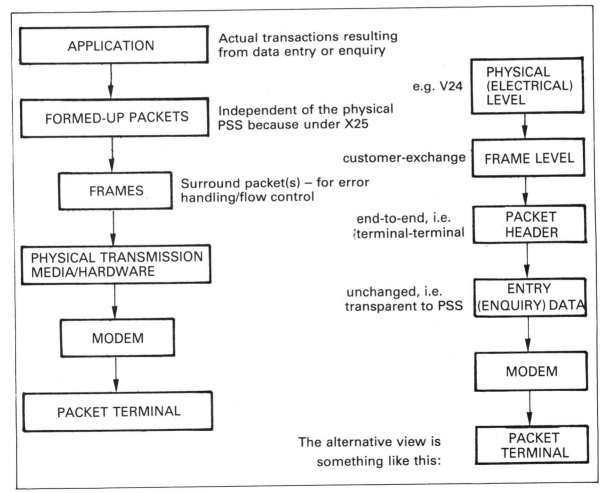

Figure 3.10   *Alternative views of packet transmission*

The framework of packet transmission

Let us look at the sending of packets from two different aspects. The first consideration is to think of the levels or layers involved. From the terminal user's point of view, we have the situation shown in Figure 3.10.

Packet format A simplified view of a packet is given in Figure 3.11.

Obviously, there must be some sophisticated software/hardware to obtain this structure from the raw data sent synchronously or asynchronously from the originating terminal. (The conversion/deconversion must take place in the PAD or in the packet terminal itself.) In addition, to allow for standardisation, the whole area of packet assembly/disassembly must be the subject of protocols. We have seen already that the 'X' series of CCITT recommendations relates to data networks. X3 specifies what a PAD should be able to do in a public data network (PDN). X25 specifies the link between the DTE, e.g. a terminal, and a DTE, e.g. a modem, for packet-mode operation on public data networks. X3, X28 and X29 relate to the link between terminals and a PAD in a packet-switched PDN. (The slang name for these three is 'triple-X'.)

Physical level This involves the actual electrical signals, etc., that are involved in the connection between terminal and modem. (We have already mentioned V24 in this context.)

Frame level At the frame level, the area of concern is with error handling and control of the flow of data. During transmission there may be random electrical/electronic faults which may turn bits on or off, or a whole burst of interference may arise which affects several packets. The receiver must be able to reject packets when this occurs even if it means carrying an overhead in 'redundant' data.

There are various techniques for 'redundancy checking' as we have already discussed and any of these may be applied, especially the use of a checksum. This is attached to a block of data and if the receiver does not compute the same checksum, a request to retransmit will be issued.

This checking is carried out on a 'hop-to-hop' basis in that, when the data is passed through a PSS, it will involve several packet switches or exchanges and each of these must be able to request retransmission after a certain time period (called a 'timeout') or if a negative rather than a positive acknowledgement is received from the next one down the line.

Flow control includes the assembly and disassembly of messages, the routes selected and the general control of congestion during heavy usage.

The data message to be sent is broken up into packets which could go by differing routes to the receiver and there must be a mechanism for sequence-numbering so that the receiver knows when an 'out-of-sequence' occurs. This is particularly important in very high speed networks. The technique whereby the next packet is sent only after the previous has been confirmed and a packet is retransmitted if necessary, is called *Automatic Repeat Request (ARQ)*. It can be thought of as working in a half-duplex mode in that the sender only sends after an acknowledgement. For high-volume transmission we cannot afford to wait for the acknowledgement and many packets must be sent. This means that the system must buffer the packets as well as sequence them. It also means that the receiver must be able to inform the sender which packet it needs to have retransmitted (called 'selective' ARQ).

One of the bases of this level of control is *HDLC (High-Level Data Link Control)* as specified by ISO. The packets of information/data are framed as shown in Figure 3.12.

The *flag* denotes beginning and end of frame. The *address* tells whether the packets contained in the

From address	To address	Packet sequence number	Packet contents	Error detection facility

Figure 3.11 *Standard 'packet' format*

Flag	Address	Control	Information/ data packets	FCS	Flag

Figure 3.12 *HDLC 'packet' format*

frame hold data, or command information such as that to acknowledge or request retransmission). The *control block* is used to ensure that retransmission occurs, handle the packet sequence number, etc.

The *information/data* block either contains packet headers and packet data or command packets. *FCS* is the 'Frame Check Sequence' which is the checksum data referred to.

The structure and content of the frames must allow for an 'idle link' condition when nothing is happening. The 'supervisory' frame is continually sent to a terminal to inform it that the link is up.

Packet level Here we are concerned with the standards and protocols that relate, among other things, to the actual addresses of sender and receiver (NUI and NUA). In a PSS, a *Network User Address (NUA)* is assigned to a user of the service, i.e. the dataline. By international agreement, this consists of 12 mandatory digits with two optional extra ones for use within the user's own network. The first four digits are called the *Data Network Identification Code (DNIC)*. (For Switchstream, this is 2342.)

Other information relates to the type of packet and the way it is being sent. For example, it is quite likely that part of the link may be operating in multiplex mode and it will be necessary to indicate whether a particular connection is both ways or in one or the other direction only.

Yet other parts are used to indicate non-standard options and to set up call connect and disconnect procedures.

There is obviously very much more to packet-switching but the forgoing will give you some idea of what is involved.

BT Packet Switchstream facilities

For the system, a *call* is a number of packets sent as a batch. The system uses the concept of a *logical channel* (or *virtual circuit*) over the datalines which means that several user calls can be multiplexed and share the 'line' by packets being interleaved.

There are two kinds of call referred to as 'datacall' (maximum length 128 bytes) and 'minicall' (maximum length 1024 bytes). Messages are usually sent as datacalls which have three 'components':

● Call set-up which consists of 'call request' and 'call accept'.

● Data transfer.

● Call clear.

Because of the way that the datacall is structured, it is possible to use the call request component to send a minicall and the call accept to reply.

Access to the system is by packet or asynchronous character terminal along a dataline or through the PSTN (dial-up) and Figure 3.13 shows what is possible.

To access the system by character terminal, you are allocated a *NUI (Network User Identity)* which is recognised by the PAD (not the NUA

		Datacall	Minicall
Dataline connection	Packet terminal	x	x
	Character terminal	x	Receive
PSTN connection	Character terminal	x	

Figure 3.13 *PSS access routes*

mentioned above which is to identify you to other users of the system).

Access via dataline is different because your packet or character terminal is permanently connected to the PAD via the dataline and switch-on automatically links you in.

For dataline access terminals, several facilities are available, in particular 'Transfer call charge acceptance' which allows the receiving terminal to refer the charges to the sender, and the 'CUG (closed user group)' which restricts calls to and from terminals to others defined as a closed group. The CUG will be referred to again in Chapter 6 when we look at viewdata, since a CUG to all intents and purposes is the equivalent of a private viewdata system.

Variations are possible so that a CUG member can have outside access (send to users outside the group), incoming access, or both. It is also possible to receive telephone bills analysed (broken down) by member or sub-group.

Packet terminals can have 'call redirection' whereby if the dataline or terminal is out of action, the call can be re-directed to another packet terminal, which can optionally redirect one more time.

A recent addition is the *Multiline* service in which several users can share the same NUA, i.e. the same dataline, even if they are operating at different speeds, provided they link at the same PSS exchange. With this facility, the number of calls is greatly increased and if one line goes out there are still others available.

A large user may choose to install his own packet-switched network either for security/privacy reasons or because it is cheaper. Because of the acceptance of and adherence to the standards and protocols associated with packet switching, these private networks can still be linked with the public switched and other networks, such as Telex.

Other X-stream facilities

Four other facilities resulting from the application

of digital technology can be leased fom BT.

Kilostream offers point-to-point digital transmission between city centres at speeds from 2400 to 64,000 bps. Typical recommended applications are high-volume speech, high-speed electronic mail, fast fax and credit verification.

A 'structured service' is provided for rates up to 48 kbps which means that an extra two bits are added for each six bits to be transmitted. They are used for fault diagnosis and monitoring between the network itself and the *network termination unit (NTU)*, i.e. the interface between the user and the circuit attaching him to the network. The 64-kbps service, which does not involve the extra bits, is referred to as an 'unstructured service'.

The low-speed service can be multiplexed with the use of several different statmuxes, the medium-speed service lends itself to most muxes while special high-speed muxes (TDM) are needed to make possible the use of voice and different types of data over a particular Kilostream circuit.

Master Systems (Data Products) Ltd claim that their *Mastermux IV* can handle up to 60 separate data channels on a single 64-kbps Kilostream bearer with any mix of terminals with different speeds and different protocols.

Megastream provides extremely high volume voice or data transmission, separately or mixed together. It is normally used to provide 2-Mbps transfer but can be specially adapted to speeds of up to 140 Mbps. With suitable multiplexing, a combination of voice and data can be handled, such as 30 voice channels at 64 Kbps or a higher level of multiplexing by reducing the 64-kbps bandwidth further, e.g. 4×16 kbps. It finds application where enormous quantities of data need to be transferred such as with image transmission (Chapter 5), video-conferencing or engineering applications such as geological/oil surveying. It also finds application in linking multiple SPC exchanges together to allow multiple concurrent speech conversations.

Satstream offers linkage to users with ultra-large volume or long-distance requirements to other users by satellite, transmission to these destinations being via roof-mounted microwave dishes.

The London Overlay is another microwave-based service, for City of London users such as stockbrokers, banks and financial information providers. It gives an option of 6 or 30 high-quality channels and can handle up to 2 Mbps.

BT claim that they can link you in to another overlay user in 24 hours. Its main advantage is that calls are transmitted over its circuits extremely quickly to the receiver's serving exchange, thence to his terminals by usual lines. The effect is that users can acheive a very high speed, high quality service apparently with their normal equipment.

Multistream

This is a fairly new BT facility that gives the asynchronous, character terminal user an efficient data network based on packet-switching facilities via the PSS. It can be utilised by a single user wanting convenient access to computers, databases and videotex, but it has built into it a series of user addresses and passwords that allow a user to offer a kind of limited, private data network.

The PSS has its own, internationally recognised error-handling procedures, but for the non-packet terminal user, the linkage to PSS is usually via dial-up through the local PSTN exchange into a Multistream PAD and these safety measures do not apply from terminal to PAD.

Multistream is designed to provide convenient 'datacall' access to computer and videotex services connected to PSS, at the same time covering the link to PSS with the same safety measures. The main features are *EPAD* and *VPAD* which can be thought of as data transfer protocols for general data and videotex access, respectively. EPAD is mainly involved with error-free data transfer. VPAD is also involved with the different formats associated with videotex services.

The user can have direct, 'autocall' access to a particular service that is used frequently, or menu selection to a range of services, or 'open access' to any PSS-based service. As with direct PSS usage, it is possible to ask for analysed billings, reversed charges, closed groups, etc. For convenience, once a datacall is finished, a further service can be called up without having to break the link and re-dial.

PAD control parameters geared to CCITT recommendation X3 are used as standard, but it is possible for a user to make use of extended services which make use of non-standard parameters. In other words, although most services will employ standard packet assembly/disassembly procedures, EPAD allows a user to specify a 'profile' of parameters to be associated with calls.

Integrated services

Although we have still to cover electronic mail and videotex, this is a good time to rationalise what we have already touched on, with reference to integrated digital services.

Because of time frames, volumes, costs, lack of standards and the different rates at which they have developed, voice, data and text have different requirements in switching and transmission. Until relatively recently all traffic was carried along the analogue PSTN, except Telegraph and Telex, which had their own networks and, as we have seen, analogue transmission along narrow frequency bands is far from satisfactory for data.

Another factor is the fact that with telephone, people expect to be able to link up with anyone else who has a phone anywhere in the world. This has been acheived because telephone authorities and common carriers have conformed to certain standards or have introduced interfacing equipment.

With the development of digital transmission techniques and their network provision by BT and Mercury, and the support of companies like Plessey, Northern Telecom and GEC in providing digital exchanges and other hardware as well as digital networks themselves, it would have been expected that by now, all facilities would be linked together.

This universal 'ultra-wide network' has not really happened with data transmission for several reasons, mainly because hardware and software from different suppliers for different applications are not compatible. This is hardly surprising.

Data access started off with the remote terminal

being sited within the same building as the computer, then by slowish-speed telephone link to a company computer. (The author was with GKN in 1967 when they installed an IBM data transmission link between London and the North). At the same time, connection might be to a service company computer. (The author was making use of Honeywell Time-Sharing Ltd – now Geisco – from Acton Technical college, over 14 years ago).

In the early 1970s, the terminal was either totally compatible with the computer because it was a peripheral of the computer, or was limited to something like the ASR33, which was already in use in Telex and for which the limited computer services available had already been tailored.

More recently, particularly with the advent over the last couple of years of the cheap personal computer and, since liberalisation, the cheap modem, many different users want to access many different services. Also, as we saw earlier when looking at protocols, the larger suppliers of computer hardware and services went ahead with their own standards (although more recently, there have been valient efforts to come closer such as in the area of terminal emulation).

Telephone service requirements have become more demanding and computer-controlled techniques have been employed in switching and transmission. As a parallel process, users want remote access to other users in the same company, their own and other people's computers and to all the electronic mail services. It was inevitable that ideas have evolved for an integrated approach in which access to all of these services can be integrated into an efficient, reliable, cost-effective service that:

● Offers all of the services wanted – electronic mail, telex, videotex, 'photo videotex' (high quality for photographic reproduction), etc.

● Handles voice and data equally easily.

● Does not need special gateways to link one service to another.

● Has the same terminal hardware for every service.

The hope is that eventually, everyone will have access to an Integrated Switched Digital Network.

CCITT is looking into a set of basic standards covering concepts, principles, methods and terminology for digital transmission. It is called the *I-series* (*cf.* the *X* series of recommendations for packet-switching) and is broadly structured to deal with:

● Service aspects.

● Network aspects.

● User–network interfacing.

● Network–network interfacing.

● Service maintenance aspects.

Various other recommendations will follow relating to specific existing and future networks and network components.

ISDN has many advantages in the flexibility and reliability of the services it will offer, especially for the small user. The large user can put in the services required from his network as and when he wants them. The small user can only wait for them to be made available, but to be able to rely 'on tap', on all these facilities, will mean a lot of users will have access to them without the investment needed at present.

The international nature of the standards involved means that suppliers can go into new markets knowing that there will be an increasing reliance on and adherence to standards and recommendations but at the same time, product differentiation will become increasingly more difficult. Furthermore, the user will have a choice because all suppliers will have to conform if they are to compete, which means that they will not have to rely on one supplier who in the past, has dominated the market and hence dictated terms to his customers.

Perhaps the biggest advantages for all lie in the ability to plan on the long-term knowing that there will be more standardisation than has been the case in the past, that services will be reliable and available from many different sources. The basic user/network interface at present is not geared to digital technology since over 60% of local lines are based on copper cable and it remains necessary to convince the consumer that the cost to him of laying on the hardware facilities to back up the transfer to ISDN will be more than outweighed by the convenience and cost-effectiveness of services that will be supported. Much work is being done on digital switches and PABXs and it would seem that much more must be done at the caller end.

Mercury have been looking very seriously at this in their provisions for local linkage in the City of London with underground cabling.

Integrated digital access

At the caller end, access to the digital network is called *Integrated Digital Access (IDA)* and by converting the telephone and the connection to the network, it will be possible to have *single-line* IDA on the basic twisted-pair line.

The CCITT recommendations for IDA are likely to be implemented according to these schemes:

I420

2 64-kbps B channels
1 16-kbps D channel

I421 (Europe)

30 64-kbps B channels
1 64-kbps D channel

I421 (USA)

23 64-kbps B channels
1 64-kbps D channel or

24 64-kbps B channels and other signalling

BT have a pilot ISDN in operation which from two-wire connection to the exchange gives a 64-kbps speech/data B channel and an 8-kbps data B channel with a second 8-kbps D channel for signalling. The signalling is carried out using a method called DASS. It is geared to a new City of London, System X exchange and currently covers customers in Southern England. An extension will involve three other exchanges in large cities.

An alternative is to have *multi-line* IDA with up to 30 independently switchable speech/data connections, time-multiplexed from a digital PABX to the exchange with DASS signalling for all of these multiplexed onto a separate channel. This is based on I421.

Many countries are looking at ISDN and Japan has a network service which is as advanced as that of BT.

Mercury already offer a facility applying I420 providing switched access to the network facilities. The D channel, using a packet format at 16 kbps, will be used for signalling facilities for the B channels but will have slack capacity that can be used for slow devices.

The degree of standardisation that comes with signalling techniques like DPNSS and DASS means that it should not be too difficult for companies with digital exchanges and sufficient forward-thinking to become this advanced, to progress quite economically into a public ISDN once it appears.

To date, there have been quite a few private digital networks installed. BT in particular have put in systems with DPNSS for customers such as British Rail and the Bank of America. GEC Reliance claim that their private network installed for Unilever is the largest in the UK controlled with these signalling methods.

At present, ISDN is still a theoretical concept rather than a working national/international system although BT would say that their implementation is a full if provisional offering. Many public and private organisations and companies are frantically working towards a working implementation notwithstanding that a general regulatory framework is not really likely to be available until 1988. In addition, consideration has still to be given to the various tariffs that will be specified for charging. It is to be expected that the charging of digital services to the consumer must change from being based on rates for particular services over to a volume charge, i.e. based on the number of bits sent/received, otherwise life will be very difficult.

Finally, to give an idea of how things might go, BT are planning to introduce a set of PBX-like facilities within the digital network called *CENTREX*. The system, once implemented, will allow customers to use their exchange telephone lines exactly as if they were extensions on a sophisticated PABX. A future extension of this will be *Virtual Private Networking (VPN)*. Users will connect terminal PABXs to the VPN acting to them as a CUG and establish the equivalent of a nation-wide private network Mercury have already introduced their equivalent.

Chapter Four

Telephone and telephone-linked facilities

The previous two chapters were concerned with the technology behind communications. The next two chapters will deal with data and text, so this one is really devoted to voice in the sense that it mainly relates to the use of telephones themselves. It looks at the facilities available to the user from a modern 'keyphone' or 'featurephone' and the exchange itself, and looks at some products and facilities available that make life easier for the telephone user, such as call management systems. Then we will deal with mobile telephone and touch on the telephone as a mobile data entry terminal.

To meet the demands of modern business, a very high degree of 'intelligence' is needed from a telephone system, either provided from the computer within the modern exchange or the microprocessor fitted into the telephone. We can consider this at two levels. The first is in the efficiency and quality of transmission, which is obviously a function of hardware design and the network providing the transmission paths. The second level is made up from all the services and facilities at the user level. Facilities can be quite simple, like an automatic 'redial last call' button or a visual display of what you are dialling, or more complex like the ability to be informed, when making a call, that there is another coming in for your extension. Some of the 'intelligence' comes from the phone itself, some from the PABX and some from the exchange itself. The following is a review of many of the facilities that are available, bearing in mind that however many are provided, the user will always think of something else.

Telephone user facilities

Internal/external use Some systems are intended for internal use, particularly on large sites, while other applications will obviously require access to outside calls. It is not easy at present to combine the two.

PABX/direct line Although most calls will be handled by the main exchange, there is often a requirement for one or more extensions to have private lines.

Short-code dialling from directory This is becoming almost commonplace now. The user can store a range of often-used numbers and give each one a 'short' code consisting of one to three digits depending on the power of the system. The equipment is programmed to dial out the number from the entered coded version. With the simpler featurephones, a limited form of the facility may be obtained by programming one of several separate keys so that it is dedicated to a particular number. More sophisticated systems allow for the storage of 10 to 500 telephone numbers. Don't forget, by the way, that this facility is also needed when you are making use of telex and fax, where often-used numbers can be similarly stored. This facility is also handy for corporate use when switchboards are linked by private line. An organisation may have several sites too widely separated for local 'internal' connection. A typical example might be the payroll department which is likely to be phoned all day with queries. The BT

Monarch system prefixes short-code groups with '**' and if such a system is installed, anyone wanting payroll could just dial something like **1 and **2 for personnel, etc.

With large systems, a particular user can have his own personal set of short code-accessible numbers as well as those available to all users.

The BT *Tonto* or ICL *One-Per-Desk (OPD)*, (basically the same machine) is a highly intelligent telephone/computer terminal which allows for several hundred telephone numbers together with names and addresses to be stored in its internal memory. The short code can consist of up to three characters such as BAN for the bank. Figure 4.1 shows the short code 'tim' assigned to a Mr Tim Humphreys.

The machine can also be used to dial out directly to computer services such as viewdata. Figure 4.2 shows a directory entry set up for one of the Prestel centres, the short code assigned being 'pre'.

When a short code is used to dial out, the bottom part of the screen display shows the number dialled.

The OPD/Tonto will be discussed in more detail later on.

Repeat last number   Either by a designated short code or a special function key, most systems allow the last number to be 'saved' and redialled. The BT Merlin DX allows this as follows. Dial the number, if engaged, press the 'recall' key and then key in#4 to store the call. On a later occasion, to redial, you enter #7. Making other calls has no effect on the stored number. It is removed either by a successful dial or storing another number. Note that with a powerful exchange like this, many different users can use facilities like this quite independently; there is no danger of you redialling someone else's last number.

Repeat dial   If a number is engaged, the system can either keep trying until successful, or can try several times at pre-determined intervals before giving up or returning to you.

Call barring   An obvious need in cost control, it is often possible to prevent certain calls from an extension, such as all outgoing calls or all inter-

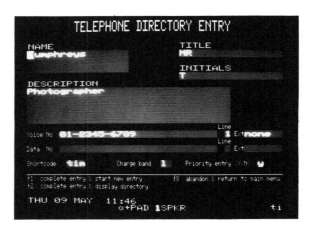

Figure 4.1   *Setting up a directory entry including short code*

Figure 4.2   *A computer services directory entry*

national calls. The reverse direction can also be arranged, whereby certain extensions will not be able to connect with a nominated extension.

Call-back   If an internal extension is busy, you can enter a code which signals that the extension is to be tried again as soon as it is free.

Coded ringing   When your extension rings, an internal call will sound differently from an external call which may be again different from a call-back call. Yet another possibility is a 'call waiting' sound when you are in the middle of a call, reminding you that an outside call has come in.

Conference call   This can allow a number of

extensions to be linked with an incoming call so that a conference can be held.

Call hold　When dealing with an outside call, you may want to hold it temporarily while speaking to an extension or make another outside call. You may be able to shuffle betwen the two calls.

Call park　This is a variation on call hold where you can hold a call on one extension and then pick it up from another one.

Camp on　When an extension is busy (such as when a secretary has picked up a call intended for the boss), the system will connect the secretary as soon as his line is free.

Direct dial　An extension can have direct line access as well as being accessible through the PABX.

Free call pickup　If an extension is heard ringing, you can transfer the call to the phone nearest to you.

Call diversion　This can have a number of different levels with calls being picked up and diverted to another extension or to the operator:

● *Do not disturb*　All calls are diverted, either to another extension or to the operator.

● *Diversion on busy/no reply*　If you are engaged or away, you can specify which extension the call is to be passed to.

● *Selective diversion*　This allows calls to be diverted on certain conditions, such as all external calls, no reply after five rings, etc.

● *Follow me*　Literally; as you move round the building, you can arrange for calls to be diverted to the next extension you expect to be at.

Grouping　A group of extensions can be logically linked together for the following:

● *Even distribution*　Calls to the extension phones can be spread out fairly so that after receiving a call a particular extension will not be bothered again until all the others in the group have received a call.

● *Hunting*　The system will pass an incoming direct or operator-collected call to a number of extensions in a pre-determined sequence until a free one is found.

● *'Internal' connection*　Members of the group may all be dialled with their own two-digit 'short' code and the ringing tone from a group extension sounds differently from one outside the group.

Miscellaneous facilities are as follows:

● *'Executive' over-ride*　Users can be given special privilege so that they can break into another call or over-ride a 'do not disturb' or a call diversion.

● *Special privacy call*　Super-special privilege where a private call cannot be interrupted, even by another executive.

● *Music on hold*　When an outside call is waiting for an extension to clear, appropriate soothing music can be plugged in.

● *Charge advice*　Entering a code before or during a call signals the system to inform you of the charge for the call made.

● *Reminder*　A timed reminder call with an option to give a recorded message.

● *Answering machine*　Gives out a recorded message when you are away and collects messages from your callers.

● *Birthday call*　A message can be recorded and sent at a specified date and time.

● *LED display*　To show call progress, number dialled and facilities accessed.

● *Calculator/clock*　'High-level' conveniences, but quite handy.

● *On-hook dialling*　Being able to ring out and only pick up the handset when a connection has been made.

● *Speaker*　Being able to switch from handset to loudspeaker/remote microphone for 'hands-off' operation.

Featurephones, telephone systems and exchanges

Some of the facilities mentioned in the previous section obviously relate to an operator while others are designed to remove the need for an operator except for very busy exchanges or those where customer call servicing is complicated.

A vast range of facilities are provided by a System X local exchange because of its digital signalling and if you have local System X service, you have access to the *Star* services based on the keypad 0–9 + * + # and automatic voice guidance to explain as you do it. These include:

- Repeat last.

- Short code dialling.

- Alarm call.

- Call waiting.

- Call diversion (directed, engaged, no reply).

- Call barring.

- Three-way calls.

- Message call.

- Call charge advice.

A number of extra facilities are available from System X (even if the linkage is to a non-digital exchange) such as automatic reverse charging, credit-card calling and selective accounting (leading to analysed phone bills). A particularly interesting facility for companies that deal with telephone sales is the 'universal access' telephone number. A single number can be nationally advertised and calls to the number from wherever they originate are routed to the local branch that deals with the caller. BT have recently installed 'front-end processor' equipment at one or two local non-System X exchanges to make them act as if they were.

In the area of exchanges, telephones, etc., the terms used are really quite bewildering. You will hear of 'telephone systems' which can be a key system or a medium PABX, 'communication systems', 'call distribution systems', etc.

The term *featurephone* is used to describe a telephone that has keys to access the facilities from the exchange, while *key system*, or *key telephone system (KTS)* is used to describe automatic, mini-exchanges, typically less than 30 lines in capacity. Featurephones have microprocessor logic that can handle jobs like converting a short code into a number, or expanding single-key operations into exchange signals. For example, with the BT *Monarch*, hitting the 'divert key', followed by the extension number to divert to, has exactly the same effect as entering *51# and the extension.

With KTS, the idea is that anyone can pick up calls and transfer them or invoke some of the other facilities although above a certain size, an operator is likely to be needed. The BT Merlin *Herald*, which they call a '*phone system*', can either be controlled by an operator, or one of three phones can be nominated as the dominant one, perhaps handled by a receptionist.

However, other people would say that the essential difference between a KTS and a PABX is that the PABX needs an operator while the KTS does not. If you look at the list of facilities above, it should be fairly clear which ones can only be controlled by the operator. In addition, a KTS can often be 'piggybacked' onto a PABX to make use of its more extensive facilities both for internal and external use.

Figure 4.3 *BT Monarch IT440*

The main aspects of a featurephone are that it relates to the system for which it was designed and cannot usually be replaced by a standard telephone since, in general, a featurephone requires more than two wires for communication and signalling. There are, however, some *hybrid* featurephones which can plug in directly to the phone socket and into the system, such as BT's new *TX72* which is the featurephone for the *Monarch IT440* 'big switch'.

Similarly, the *TX25* featurephone is intended for the *Octara 32* key system, and the *TX58* for the *Pentara* key system. The *TX14* will serve *Viceroy*, *Regent* and *Kinsman* (all small-to-medium systems) as well as the big SX2000, since all are manufactured by the Mitel Corporation.

The 'big switches' like the *Monarch*, *BTex*, *DX* and *SX2000* are referred to as communications systems because they can act as voice/text/data switches and, with extension hardware, can handle all kinds of terminal: word processor, telex, teletex, viewdata, etc.

There is no comprehensive British Standard for all aspects of key systems at present, but the Department of Trade and Industry has produced two sets of requirements named *SCRAP (small call routing apparatus) I* and *SCRAP II*.

Figure 4.4   *BT Merlin TX72 featurephone*

The earlier standard related to systems not intended for external (PSTN) use so could cover the telephones themselves, such complete systems being called 'black boxes' and marketed by Ansafone, Intercom, Plessey, GEC, etc. These systems have a limit to extension line lengths of a few hundred metres which makes them satisfactory for most local sites.

Exchanges themselves have undergone a total revolution, as you saw in the previous chapter,

Figure 4.5   *Merlin Viceroy and Kinsman consoles and the TX14 featurephone*

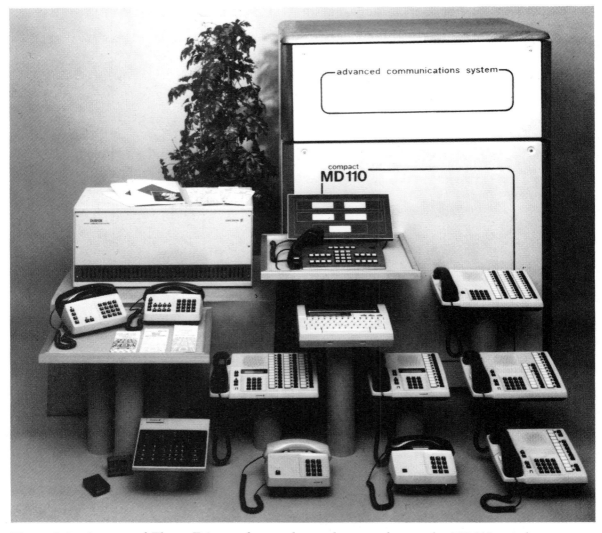

Figure 4.6 *A range of Thorn Ericsson featurephones that complement the MD111 switch*

whereas the telephone itself, although changing dramatically in shape and size throughout this century, was basically the same machine until the advent of MF dialling. Now featurephones look almost as complicated as switchboards.

Many features are now provided by modern digital exchanges, whether a large one like an *IDX* or *DMS250* or a small one such as supplied by companies including BT, STC, Thorn and many others. The degree of sophistication available is probably more than a match for the average user, because of the basic lack of 'user-friendliness'.

After all, if you want sophisticated facilities, you must expect to spend some time learning how to get them. The suppliers of featurephones or key systems have put products on the market that can link in with the PABX, or stand alone and facilities are usually accessible from the various buttons or keys fitted on the machine.

When choosing a system you must consider the phsychological effect of 'making' the user do his own work rather than allowing an operator to handle the various connection problems that arise. Also, some of the facilities are very clever, but will the average user bother to learn to use more than a few of them? If the user is to be able to take advantage of all the many facilities that can be provided, one would hope that manufacturers will eventually start to employ some of the thinking that leads to user-friendliness in computer data entry.

Some of the featurephones associated with Thorn's *Compact MD110* switch have user-programmeable keys so that quite advanced features can be obtained with a single key. The Mitel *Superset 4* makes clever use of an integral display which shows some of the features available with the keys that give the facility, lined up under the display. It has a separate set of keys which can be set for its other facilities.

Notwithstanding these two examples, there is an incredibly wide range of key-oriented systems to choose from.

The Autophone *Delta* is a single featurephone which has a display, a 10-call memory and six special-function keys including recall to PABX, save/cancel memory entry and call timer.

A system made by NEC and distributed after customising by Ansafone is the *E616*. This is a key system that can cope with up to six lines with 16 extensions. It has 20 memory dial keys and provision for a further 40 'speed dial' numbers. There is a display showing which lines are selected and their status and, when an extension is lifted, if a

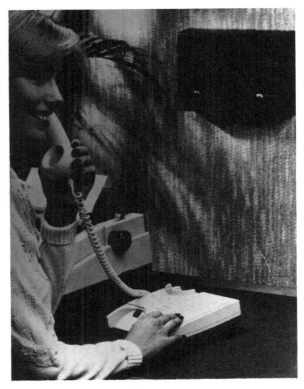

Figure 4.8 *Philips KBX6 telephone system*

line is available it is automatically assigned to the extension.

Philips Business Systems sell a neat telephone system they call the *KBX6*. The unit is wall-mounted and supports many of the useful facilities listed above. A range of featurephones go with the KBX6 including their *KT1* and *KT4*, both of which can be designated as a 'hot line' up to 5 km from the central control unit. When the hot line handset is picked up, a secretary's terminal rings for 15 seconds, then an exchange line is automatically connected.

STC produce *SDX*, a small digital exchange that can deal with up to 40 lines and is wall-mounted. It has an output that can be connected to a printer for the production of usage reports. A range of four featurephones support the exchange, most of which have a neat LCD to show which features are in use.

As we have already seen, BT supply a very extensive range of systems and featurephones.

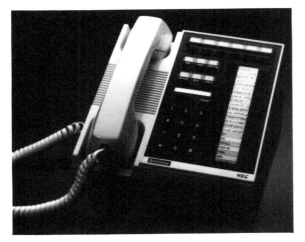

Figure 4.7 *Ansafone E616 featurephone*

Figure 4.9 *Philips KT4 featurephone*

The *Octara 32* will handle 10 lines and 32 extensions. The unit is wall-mounted and offers three different featurephones.

The *Pentara* will handle up to 16 lines and 76 extensions.

The *TX14* already mentioned can show the status of other TX14s and has 'secretary' facilities such as camp on busy, message to the executive's display and break-in. The phone can also tell you who is calling in.

Viceroy can handle eight lines and 16 extensions. Its facilities make it useful in hotel applications. Amongst other facilities, it can produce separate billing information and has an automatic 'wake-up' whereby an extension can have an alarm given at a pre-set time. If not answered, the call may be repeated twice more.

The *Regent* can handle 28 lines and 126 extensions or for special applications, 12 lines and 182 extensions, up to 64 of which can be *TX14* featurephones. This is also recommended in

hotels. Extension numbers can be the same as room numbers. Different charge rates or supplements can be associated with different extensions, printed bills can be obtained from the system and it features a 'baby listen' facility. By coding the extension when you go out, you can ring back from another and set the extension microphone so that you can listen for a baby crying.

Linkline 0800/0345

Although not a digital service at present, this seems as good a place as any to mention *Linkline*. This is a rather different BT service aimed at companies concerned with telephone sales/promotion and sales enquiries. To make an enquiry from anywhere in the UK, the user dials a number consisting of the code 0800, followed by a unique six-digit number for direct connection to the company being called. It gives free access to the company, with the Linkline subscriber picking up the cost of the call.

The service uses the same transmission paths and techniques of the PSTN, but has its own regional switching centres.

Figure 4.10 *Merlin Octara 32 digital key system with 3 featurephones*

0800 222444 will give you the contact for BT's list of brochures on networking and advanced telecommunications.

A variation is to link to the subscriber by dialling 0345. For this, the caller pays the local call charge and the subscriber pays the difference between this and the full Linkline charge. A call is passed to the nearest Linkline exchange and is routed to the switching centre for the called company's area. For an 0800 call, the standard metering of the caller is shut off and for 0345 calls, it is set to the standard local call rate. Extra call logging and other equipment is built in so that the called company can be charged for calls and so that the bill can be analysed by the caller.

A similar '800' service has been available in the USA for over 10 years and the 'International 800' extension to Linkline from British Telecom International, allows customers access to the USA and to other countries.

BT are planning for a digital network which will provide the additional facilities expected these days such as inserting announcements, call diversions and queuing.

Both the national and international facilities are catching on and when linked with credit card sales make an extremely interesting and useful application on the telephone. Barclaycard have recently taken on 0345 555555 to provide local rate credit authorisation from their Northampton centre.

Voice messaging

Voicebank

The most powerful example of this is a sophisticated voice 'mailbox' facility offered by BT called *Voicebank*. In essence, it allows you to record and receive voice messages. You use a special Voicebank telephone number and after confirmation of your security code, you can record a message or pick up messages stored for you. You can assign a further level of security with an extra security code.

The system is supported with a keypad which you use to dial in and select facilities. As you enter

instructions, recorded voice guidance is obtained. Voicebank will confirm that a message left has been received and if you have a bleeper, it will let you know that a message is waiting for you. At any time, up to 10 users could be recording messages for you and up to 48 one-minute messages can be held, for up to 72 hours. If you require more capacity than this, it can be arranged that the 'overflow' goes into one or more other Voicebank mailboxes. For convenience, you can set up a directory of regularly used numbers.

Extra facilities include:

- *Retrieve-only option,* where you do not need to record messages, only receive them. In this case you will not require the keypad.

- *Advance booking* This allows you to record a call up to nine days before it is to be received.

- *Group broadcast* By making one call, you can send the message to a number of different boxes.

- *'Guest' broadcast* This allows you to leave a message for a non-Voicebank user and accept his reply.

- *Message-edit* With this, you can pick up a message from your mailbox, edit it and re-record it for someone else.

When we look at cellular telephone, you will see that Racal-Vodafone offer a simliar but more limited service for mobile telephone users.

Plessey IVM

Plessey sell a system called *Integrated Voice Messaging (IVM)* that enables you to set up your own 'voicebank' as an adjunct to a modern exchange such as their IDX (which you will recall is sold by BT as their DX). The system comes in several different sizes, the largest handling up to 500 mailboxes, 6 input ports (6 simultaneous calls) and 64 messages up to three minutes long per mailbox. Messages are stored in digitised form on a large magnetic disk (the largest system has a 160 Mb disk) and password-access to messages is by MF telephone or keypad. General facilities are rather similar to Voicebank.

Call management/logging

Call logging or call management systems are microprocessor-controlled units that plug into the PABX and make detailed records of all calls. The information can be analysed later on and acted on to detect abuse, to monitor performance or for call charging.

There are two main advantages to the use of a call logger. The first is obvious in that it points to over-use or fraudulent use of the system, thus enabling the appropriate control to be exercised. It is very convenient to know who is calling home or ringing out for the Test Match score, but it is also useful to be told that a telephone sales team is costing more than a budgeted amount. The second, and not quite so obvious, application is in determining network under-usage. In a recent investigation carried out at a Department of Trade and Industry site where there had been a request for a replacement/upgrade to an existing 1200-line exchange, it was found in fact that the whole site could be adequately serviced by 700 lines.

G A N D O F F   B O O K S   LTD

DEPARTMENT 05 ADMINISTRATION   DATE 07/06/86

EXTENSION NUMBER	TIME	NUMBER CALLED	DURATION	COST
107	11.07	048673542	00.04.10	0.830
107	11.38	473141	00.00.30	0.130
108	09.15	0708200136	00.03.34	0.650

Figure 4.11 *Call listing*

With a logger, management can continuously monitor the use of the system and be sure that it conforms to expectations. This can also be of considerable help when trying to decide on which of the range of telephone user facilities are required from the exchange for optimum performance.

Another possibility that comes from the print-out information is the ability to apportion call charges to different departments, contracts or companies going through the same switchboard. General logging services allow for a whole range of print-outs to be produced, including:

- *Call listing* A detailed report, showing what calls were made (see Figure 4.11 for an example).
- *Selective reports* Print-outs showing all calls taking more than *x* minutes or costing more than £*x* or all calls to a particular number. A report such as that shown in Figure 4.12 is possible from the *Exchequer* call logger from Datapulse Ltd.
- *Usage histograms* These can provide a neat at-a-glance view of collected information (see Figure 4.13).
- *Department totals* An example is shown in Figure 4.14.
- *Specific reports* An example is shown in Figure 4.15. Reports can be even more comprehensive such as showing where budgeted costs have been exceeded for a line, department, customer, etc., breakdown by time of day, charge band, whether local, trunk, PSS, etc.

DESIGNATED NUMBERS REPORT - G A N D O F F   B O O K S   LTD

period 21/02 - 05/03   page 1   05/03/86

designated number	called party	number of calls	cost	total duration	average duration
012823465	R. Kinge	11	10.45	00.57.14	00.05.20
027244912	C.Collier	6	3.50	00.24.00	00.03.00
72456	Midland Bank	8	3.80	00.28.00	00.03.50

Figure 4.12 *Selective reports*

S A N D O F F   B O O K S   L T D

GENERAL CALL STATISTICS - TIME TO ANSWER   05/03/86

number of calls	ringing time average	ringing time maximum	time period start	seconds *** average >>> maximum (10 20 30 40 50 60 70 80 90+)
34	13	39	9.00	***>>>>>>>>>
35	14	40	9.15	***>>>>>>>>>>
36	15	38	9.30	****>>>>>>>>
[illegible]	25	94	9.45	******>>>>>>>>>>>>>>>>>>>>>>>>>>>>
18	27	67	10.00	********>>>>>>>>>>>>>>>
27	21	35	17.00	******>>>>>
17	9	28	17.15	*>>>>>>>

Figure 4.13
Usage histograms

S A N D O F F   B O O K S   L T D

DEPARTMENT SUMMARIES   Week ending 05/03/86   page 1

Department number	Department name	Cost this week	Cost year-to-date	Budget year-to-date
01	Personnel	24.30	260.35	250.00
02	Payroll	28.50	290.50	321.00
05	Sales	190.26	3603.27	3000.00

Figure 4.14
Department totals

S A N D O F F   B O O K S   L T D

LINE USAGE REPORT   05/03/86   page 1

Line	Incoming		Outgoing		Total	
number	calls	av.duration	calls	av.duration	calls	av.duration
001	117	1.07	123	2.13	240	1.84
003	123	1.45	128	1.54	251	1.50
008	181	0.56	127	3.49	308	2.25

Figure 4.15 *Specific reports*

A number of companies supply call management systems. Norex Systems Ltd sell two models, both of which can operate with any SPC PABX that has a V24 port including BT DX, Monarch, Regent, Kinsman and Herald. The small one called *Call Check* is a self-contained unit with integral printer. A wide range of reports is possible and these can be produced after each call or at intervals.

Figure 4.16   *Ansafone Telcost 64*

The larger *Countess* holds data on up to 4000 calls with optional extension to 12,000.

AGI offer a call management system aimed at 40–120 extension systems such as Monarch, Regent and Herald. The system is based on the powerful Motorola 68000 microprocessor and comes in four models with call-record capacities ranging from 5000 to 65,000.

Dynamic Logic have announced their *Sabre 1000* and *3000* telephone management systems which are compatible with most manual and automatic PBXs. The 3000 can handle 8000 calls on up to 100 lines, trunk and extension combined.

Ansafone supply *Telcost* which like many other loggers, is geared to Regent, Herald and Kinsman and can cope with up to 1600 lines, collecting data on incoming and outgoing calls. It will generate a very wide range of reports like the examples above, either automatically or on request. Collected data can be stored on a data cartridge for historical purposes and later analysis.

Monarch Airlines, the Luton-based independent airline, installed a Telcost last year to log their 120-extension Regent, which also has 12 direct lines. Because of the nature of the business, Monarch require a 24-hour telephone service which often involves visiting clients ringing back to their offices. The logger is said to have cut a £36,000 phone bill by 30–40%, mainly by spotting areas of unauthorised use, but it also showed the company that a direct out-of-area line for London calls could save about £5000 per annum in increased efficiency.

Thorn-Ericsson produce *Sterling* and *ASDP162*, which they refer to as 'call distribution systems', aimed at large companies whose main business is telephone sales and enquiries (such as banks, credit card companies, car rental firms, etc.). The systems are mainly aimed at ensuring that calls are taken in rotation and dealt with by the correct clerk or group of clerks. At the same time, they can monitor system performance and generate many useful reports including staffing levels throughout the day, 'renegade' or abandoned calls, line utilisation and call analysis by a particular dealing agent or group of agents.

For companies that do not want the capital outlay of logging equipment, British Telecom amongst others operate a call logging consultancy service called *Tallis Consultancy*. They will connect their logger to your PABX, collect data and produce a wide range of reports on a contract basis.

Not all users need the sophistication of full logging systems and companies such as Recordacall supply single-line loggers. Their *400CL* is fitted with an LCD display and records date/time/number dialled and duration on a single line or individual extension.

British Telecom also supply loggers. The two systems are *CM7000* and *CM8000*. The machines allow logging on a personal basis so that call information is recorded with the name and department of the user. Text can be edited and formatted to any requirement with the CM8000 since it comes with an optional *WordStar* word processor

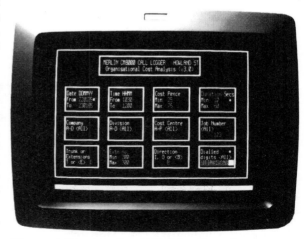

Figure 4.17   *Merlin CM8000 call analysis screen*

and the system has enough processing power and operating system control to enable text formatting and printing to proceed while logging is in progress. The two machines are equipped with VDU and reports can be pulled off at any time. The CM8000 has hard disk capacity.

The systems can be fitted to a range of BT PABXs and featurephones as follows:

Switch system	CM7000	CM8000
DX		●
BTex		●
IT440		●
Regent	●	●
Monarch 250C	●	●
Monarch Compact	●	
Pentara 100	●	
Herald 100B	●	
Kinsman 543	●	
Viceroy 243	●	

BT also produce a much larger system called *BTS123*.

Among BT's customers are Credit Factoring, who are part of the NatWest group and who use a CM8000 for call management. They produce an exception report for each of their five divisions, showing all calls costing more than £5 and lasting more than 20 minutes. Monthly reports include total divisional totals and the 50 most expensive calls.

Tonto/Qwertyphone/Displayphone

To give you some idea of the sophistication that is put into telephone systems, it is worth saying something about three products, two from BT and one from Northern Telecom. The BT Merlin *Tonto* and Northern Telecom *Displayphone* seem to offer similar facilities in many respects, but no attempt will be made to compare and contrast

Figure 4.18 *Tonto in use*

them, mainly because the author has a vested interest in Tonto (*Business Communications with the Merlin Tonto*, M. Gandoff, Century Communications, 1985), but also because the Tonto is now well-established, having been around for over a year, while the Displayphone has only been out for a few months.

The BT Merlin *Qwertyphone* is a very recent product, which although not having all the facilities of Tonto, is about a quarter of the price.

Merlin Tonto/ICL One-per-desk (OPD)

Tonto and the OPD are essentially the same machine with slightly different software, the following remarks being related to Tonto specifically.

Tonto is a highly intelligent hands-on/off telephone and a computer terminal which also has an on-board calculator and a 'messaging' facility allowing a kind of private teletex between other Tontos, as well as file transfer between Tontos or to mainframe computers.

In addition, the machine can act as a stand-alone electronic office computer and a stand-alone microcomputer, running Basic and other languages.

The machine has a QWERTY keyboard, a multifunction keypad, a one– or two-telephone line capability, integral modem, a range of compatible printers, a VDU, a large amount of RAM and, for memory back-up, two 'Microdrive' micro tape cassette drives, each of which can hold 100K characters of data. The cartridges require much more care in use and storage than traditional tape cassettes and are certainly more bother than disk drives.

Access to facilities is by means of a hierarchy of menus from the 'top-level' menu.

As a telephone, Tonto is incredibly powerful, giving short code dialling from directory and includes a facility that is beyond most other featurephones – the ability to search a directory for a name or company type. The most commonly used numbers can be displayed in a 'priority' directory.

The keypad can be used for dialling numbers in full or by short code and most of the keys are bifunctional. For example, # when shifted, becomes 'print', a special function key which signals that you want to print a 'screen dump', i.e. a print out of the current VDU screen contents (message, viewdata page, etc.).

The VDU display is divided so that the bottom part tells you the status of the machine and which particular functions you may be carrying out (it is

Figure 4.19 *A Microdrive cartridge*

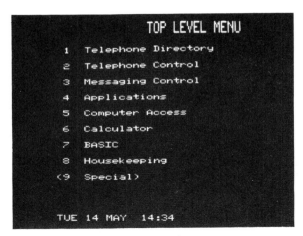

Figure 4.20 *Tonto top-level menu*

Figure 4.22 *The telephone keypad*

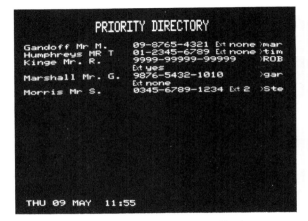

Figure 4.21 *A Priority Directory display*

Figure 4.23 *The Tonto Voice Response library*

possible to switch from one function to another and flip back again with the 'resume' key.)

Tonto can be used as an answering machine and it is possible to set up a pre-recorded message made up by keying in your selections from the words in a 'voice library'.

As a computer terminal, Tonto can access telex (via Prestel or Gold) and viewdata, and link directly to mainframes with which it is compatible. (Many of the viewdata screen shots in Chapter 6 were taken from a Tonto.)

The Tonto is sold with a package called *Xchange* which consists of four interlinking modules, a word processor called *QUILL*, a database handler

called *ARCHIVE*, a spreadsheet called *ABACUS* and a 'graphics' program called *EASEL*.

As a 'messaging' terminal, a Tonto user can use Quill to prepare text messages and while he is doing so, the machine can send messages, prepared earlier, from the 'out-tray' and simultaneously receive them in the 'in-tray'. A new facility allows for 'broadcast' messages, in which a message can be sent to a number of other Tontos.

Many options are available with Tonto as a computer terminal. To link in to different bulletin boards and viewdata facilities, you can alter the baud rate, set parity, etc., and store the information as a directory entry 'profile' for the service wanted. Frames received during computer access can be stored in memory for later printing out.

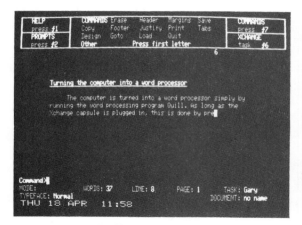

Figure 4.24 *Quill in use*

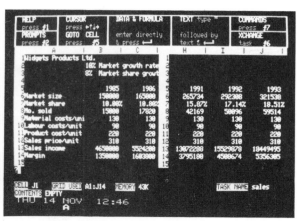

Figure 4.26 *Abacus in use*

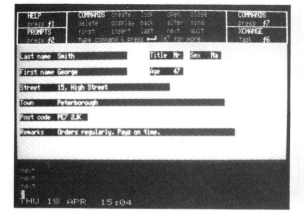

Figure 4.25 *Display of an Archived record*

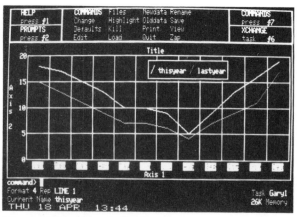

Figure 4.27 *Easel display*

Since the machine is very sophisticated in its scope, it takes a fair time to get used to its many useful facilities, but the intelligent user has a complete workstation available if wanted. Quite recently, BT have announced 3.25-inch floppy disks (720K capacity) for Tonto and it is now possible to interface the 'advanced' messaging facility with the Xchange package and viewdata files. Inter-Tonto file transfer is also supported so, for example, a telephone directory or word-processed report can now be transmitted to another Tonto user. (It is the author's strictly personal view that if these facilities had been available with Tonto right from its launch, BT/ICL might well have doubled or even trebled sales by now.)

Northern Telecom Displayphone

The Displayphone looks not unlike Tonto except that it has a pull-out keyboard, a spread-out, touch-sensitive key pad and a set of special-purpose keys. It would seem to have many of the facilities of Tonto, although the memory is more limited and it does not have the equivalent of Xchange.

BT Merlin Qwertyphone

This machine is conceptually something like the Tonto, with more keys and without the stand-alone micro facilities. BT refer to it as a desktop

Figure 4.28 *The Northern Telecom Displayphone*

Figure 4.29 *BT Merlin Qwertyphone*

terminal with a full range of alphanumeric keys plus a set of ten special-function keys, seven of which are programable for your own functions, and a telephone keypad. It is also equipped with a 4×32-character line LED display. During operation, the function of each of the nine keys around the screen is shown on the display and, as a function alters, so does the display.

The machine, like Tonto, can link up with viewdata and can handle a 250-number memory with short code dialling. A simplified version of the Tonto messaging facility is provided, with text being entered using 'memotyper' which is a simple line editor (simplified word processor). It can also act as an add-on to a PC with a special protocol that allows a PC to control the telephone functions and access stored data. In addition, it can act as a 'smart' modem and has been built to respond to Hayes commands. With the addition of a 'telex server', the machine can receive and send telexes, including those prepared off-line.

Finally, the machine can be used as a featurephone on a direct line or linked to a PBX such as DX, Monarch, Regent, Viceroy and Kinsman, and a pair of the machines can be used by manager/secretary, so that conference calls, diversion, call transfer and status enquiry are possible.

Cellular telephone

Cellular voice

As noted in Chapter 3, Racal-Vodafone and Cellnet (BT and Securicor) were given licences to act as cellular telephone operators, i.e. to provide cellular telephone networks. In its simplest form, the cellular network consists of a series of hundreds of cells or communication areas throughout the country. The maps produced by Vodafone and BT (who in effect own Cellnet), show that the cells largely cover the main motorways at present, but are being extended all the time. The Vodafone 'end-March 1986' map covers about half of Southern England, the areas around the M1/M5/M6, and Glasgow/ Edinburgh/ Dundee/Aberdeen. They claim that they will cover over 80% of the population by the middle of this year. The Cellnet map seems to cover principally the same areas with less emphasis on the West Country.

By the end of November 1985, Vodafone had over 17,000 subscribers making over a half a million calls per week and it is estimated that there could be as many as half a million cellular car phones by 1990.

The cellular network is supplied by Racal-Vodafone, Racal-Vodac handles the provision of services (with over 14,000 subscribers at mid-May 1986), and Racal-Vodata develops value-added services to the network, such as *Taxifone*, *Meterfone*, *Creditfone* and data transmission over the cellular network.

A user makes radio calls from a mobile telephone handset and as a cell boundary is reached and crossed, a central computer detects the change and switches to the transmission frequency of the next cell. Since each cell has a different transmission frequency from those surrounding it, while at the same time being of fairly low power, there is little interference from other cells, nor from other sources of radio transmission. This means that there is a great advantage to the cellular approach compared with the more conventional radio-telephone transmission because it opens up the

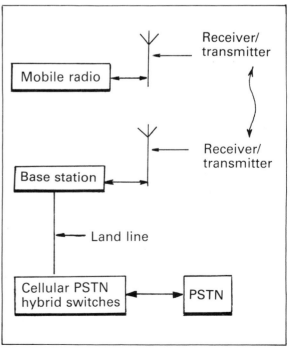

Figure 4.30   *The cellular/PSTN network interface*

Figure 4.31 *Citifone production at Racal Seaton in Devon*

possibility of using a particular frequency more than once in non-adjacent cells so less frequencies need to be assigned to a network.

Regardless of which particular cell the user is in, the central computer can pick up the transmission and link the call into the PSTN so the user is not aware that he is using a special link to the network except that the quality of the service is far superior to normal telephone or radio-telephone.

Racal have the first UK factory, in Devon, to produce BABT-approved transportables, the equipment being manufactured under licence from the Finnish company, Mobira. The 'base station' equipment, including transmitter power amplifiers and voice channel sub-systems, are produced at Racal-Carlton in Nottinghamshire.

Mobile telephones come in a variety of sizes, being divided into:

- *Mobile*, i.e. in-car; mobile in the sense that it is permanently fitted into a vehicle and the user has access from wherever he happens to be.

- *Transportable* Fitted in the car, but removable.

- *Portable* Literally for pocket use (perhaps a large pocket).

The Racal Citifone has a 40-number memory and gives many of the features listed below. An adaptor allows it to be used as a car phone as well as being completely portable. The display shows functions like call diversion and also shows the cost

Figure 4.32 *The Racal Citifone mobile cellular telephone*

Figure 4.33 *The Racal CT transportable Vodafone*

of calls made. The memory has some neat features, including the possibility of a 'partial' scan-through. If you can remember some of the numbers, the machine will display all memory-held numbers that include them.

Philips Electronics are planning to sell Citifone through their Pye Telecommunications subsidiary.

Systems are usually purchased with rechargeable battery and battery charger, and a range of accessories such as a mounting frame, handset holder and cable, and flexible or portable antenna.

Racal-Vodafone has appointed several 'service providers' to retail, install, service and bill customers. They include: Philips Cellular Radiophone (outlets including Pye), Answercall Vodafone (outlets including Dixons), Racal-Vodac (outlets including the AA) and over 20 others including Radio Rentals Vodafone and Bosch Vodafone.

The Racal-Vodafone network is backed up with what they claim is the world's largest digital cellular exchange consisting of a Thorn-Ericsson AXE10 backed up with APZ 212 processors located at Brentford in London. The network covers most of Southern England and can support 60,000 subscribers. As was mentioned earlier in the chapter, the switch allows for call waiting, conference calls and hold-for-enquiry as well as many other facilities.

BT seem to prefer to handle all the service themselves through their 'Mobile Phone Division'.

Convenience facilities

For the mobile phone user, as with conventional telephones, a number of facilities are wanted including:

● *Hands-free/on-hook dialling.*

Figure 4.34 *The Racal VM1 mobile Vodafone*

- *LCD/LED digital display.*

- *Battery low indicator.*

- *Illuminated keypad* (for night use).

- *Call barring*   To prevent all others from using your phone; also some selective protection, i.e. limited access by others, say local calls only.

- *Dialling from a directory of stored numbers.*

- *Previous number redial.*

- *Call diversion* on busy, no reply or out-of-area.

- *Call waiting warning*   While another call is in progress.

Figure 4.35
The Racal VM1 handset

- *DTMF dialling option* for tone entry/voice response systems.

- *Received call indicator*   You were out when a call came in.

- *Timed alarm*   A message can be saved and sent to you at a specified time.

- *Conference calls*.

- *Engaged redial*   When you find a number engaged, the system will attempt a redial after a time interval.

- *Charge displays*   Such as last call charge, total units used this period etc.

Figure 4.36   *BT Opal fully portable cellphone*

Figure 4.37   *BT Jade transportable*

Figure 4.38 *BT Amethyst for in-car use*

Miscellaneous developments

Racal-Vodata have recently implemented a pilot scheme in which 60 London Taxis have been fitted with mobile equipment for local, national and international calls for customer use. The meter fitted shows the current charge for a call in progress. The actual charges are 20p per unit with a minimum of 50p. The system has been approved for its safety and security by the London Public Carriage Office.

The 'London Liner' luxury coach service from London to the Midlands is now fitted with Vodafone radio telephones, which give the user credit-card access.

An interesting development that was fairly predictable is the introduction of cellular 'value-added' services such as typing and translation from calls, on-line or left as messages.

Vodafone offer a messenger service something like the BT Voicebank. A user has access to a mailbox. When busy or unattended, an announcement can be left for callers to leave their message. Up to 10, one-minute messages can be left for up to 36 hours and access is password-protected so that only you can release stored messages.

Cellular data transmission

Racal seem to have topped BT in the area of mobile data communications. There is a problem with sending data over the airwaves which is very similar to that of sending it through the PSTN – the slightest burst of noise can wreck a whole message. Racal have developed a protocol they call *Cellular Data Link Control (CDLC)* which they claim will handle many of the problems associated with cellular network transmission to the PSTN and PSS, such as signal fading due to attenuation by buildings, tunnels, etc., and Rayleigh fading which is a very complex phenomonon due to the nature of radio signals and

Figure 4.39 *A Racal cellular data terminal*

how they are reflected. These are, of course, affected by the speed of the car as well.

Another problem is 'handoff', the short break that occurs when a channel is retuned as a cell boundary is crossed. The protocol is designed to minimise these problems, by a combination of redundancy checking, ARQ (automatic repeat request) and forward error checking.

To start with, Vodafone will supply add-on units to existing kits, but they expect to be able to handle any standard interface (V24) to PC, printer, etc. They say that their modems can handle autodial/autoanswer, 1200 bps duplex operation over four-wire circuits such as PSS and the Vodafone network itself, and 1200 bps pseudo-duplex over the PSTN.

With this kind of kit it will be possible to access mainframes, viewdata and eventually telex and teletex when on the move (that is, from a car – it is hoped that users will not be using Prestel for example, while actually driving along).

Racal have announced quite recently that they will be offering private mobile radio before very long.

Trunked radio

Cellular telephone to the customer is similar to conventional telephone in that he is billed for rental, service and call charges. Using conventional two-way radio for communications suffers from the main drawback that there is no privacy or secrecy, since anyone can listen in or use the frequency, unless you are lucky enough to be able to reserve your own.

A trunked radio service is really designed for operation by small companies, perhaps servicing 500 users, compared with the many thousands for cellular telephone. Whenever a caller wants fast access, a special processor looks for a channel that is not too heavily in use and connects through that so that the user has exclusive use of the channel.

Pennine Communications are the first company to offer such a service, in the Lancashire/Cheshire area. Switching equipment manufactured by Motorola can handle the speed of channel selection and is backed up by software that deals with automatic call-back and call queuing.

Seventeen other systems are planned under what is called the radio 'Band III' services and the DTI is to award a Band III network licence to a consortium led by Pye Telecom and including

Racal, Securicor, Investors in Industry and Digital Mobile Communications.

Trunking is particularly useful for vehicles in fleets, all of whom can share the same channel and broadcast to other members of the fleet while being protected from interception from other fleets.

A final plug for Racal. Their subsidiary, Radiopage Ltd, has been granted a licence by the DTI to set up and run a national, wide-area radiopaging network.

Telephone data entry, voice response

Autophone (UK) Ltd is part of a Swiss-based group that is heavily involved with telecommunications. We have already mentioned some of their featurephone products, but they are also involved with an exciting telephone application they call *Talkback*. The idea is quite simple in concept. Instead of having to rely on a keyboard or other terminal (bar-code reader, etc.), use the telephone itself for *data* entry. Obviously, the telephone keys limit the number of characters to 12, i.e. 0-9, * and #, although there is no reason in principle why alphabetic data could not be entered using a code based on groups of keys; for example, #01 is 'A', #11 is 'J', etc. This would be rather clumsy, would lead to many errors and, most important of all, would detract from the essential simplicity of the idea, which is to make data entry as easy and error-free as possible for the user.

The system operates as follows. The user designs a dialogue appropriate to the application and arranges to have a voice library stored at the central computer. When data is to be entered, the computer is accessed by sending the same MF tones that are used in dialling (see Chapter 3), so no modem is needed. Data can be sent from any telephone, even if it does not have MF, because entry is carried out with a *tonepad*. Figure 4.40 shows Autophone's tone dialler which measures 12×6 cm and is PP3 battery-powered.

The pad is held next to the handset mouthpiece and, as keys are pressed, the unit generates audible tones which are sent down the line just like speech. If the phone being used has the MF

Figure 4.40 *Autophone data entry Tonepad*

facility, its own dial or keypad can be used. The nature of the tones is such that unlike most other forms of data transmission, the data is actually sent in decimal instead of binary and the absense of a modem means that literally any telephone can be used as the sending 'terminal'.

Autophone offer a typical banking demonstration dialogue as shown in Figure 4.41.

As well as the basic advantages of this kind of service, extensions are possible. For example, credit can be transferred to often-used accounts such as public utilities, etc. The dialogue allows you to assign a code number to each creditor and when you want to make a payment, you select the relevant menu option, supply the code number and then the credit to be transferred. With appropriate password control, the system could be used directly for database update and enquiry as well as for general transaction entry.

The Talkback hardware consists of the Talkback processor, interfaces for several hosts, if needed

```
                        AUTOPHON

                BANK-FROM-HOME DEMONSTRATION

              (available 0930 - 1730 hrs Mon - Fri)

    TELEPHONE NO:           0252 839207 or 839324

    WHEN THE SYSTEM ASKS:        ENTER:        FOR:

    PLEASE ENTER ACCOUNT NUMBER   12345678#

    PLEASE ENTER PERSONAL
    IDENTIFICATION                9439#

    PLEASE ENTER SERVICE CODE     01#           Statement request
                              or  02#           Cheque book request
                              or  03#           Bill payment
                              or  04#           Funds transfer

    PLEASE ENTER MERCHANT NUMBER  1357#         Barclaycard
                              or  3579#         American Express
    etc. etc.
```

Figure 4.41 *Autophone
backup demonstration*

(including a packet switched interface), the telephone line interface and the vocabulary storage.

The processor can pass data directly to the host via the interface, but for security back-up, transactions can be stored on magnetic tape. Similarly, there may be applications where enquiries need to be serviced immediately and the integral tape drives allow transaction entry to be delayed for later batch processing on the mainframe. The size of the Talkback unit depends on the number of hosts to be serviced, the number of telephone lines connected and the size of the voice vocabulary, and is designed around a rack system so that modules can be added for extension.

There is no reason why MF data entry cannot be used to support data entry and enquiry from other terminals so a viewdata terminal could present a selection menu or list of products and place orders through the telephone or tonepad. The communications processor can accept a mix of data types and convert them to a standard form. In the event of problems, the system will refer the 'data caller' to a human operator so that advice can be offered before returning him to the system.

Up to 19 vocabulary boards can be held in the system, with each board capable of holding up to 32 seconds of continuous speech – about 45 words. The speech is studio-recorded to order, digitised and stored in ROM although direct input from microphone or tape can be arranged. Since the speech is stored as recorded, any language can be used.

Voice response systems are used very heavily in the US motor industry for vehicle and spares location and ordering. They also have extensive application in mail-ordering, credit-card checking, airline enquiry and reservation, freight movement tracking (the data call says that goods have reached location x at time y) and shop-floor data collection.

World-wide, over 5 million customers carry out 24-hour home banking and Autophone are negotiating with clearing banks to try and have their system adopted in the UK.

Figure 4.42 *The Talkback unit with two integral magnetic tape drives*

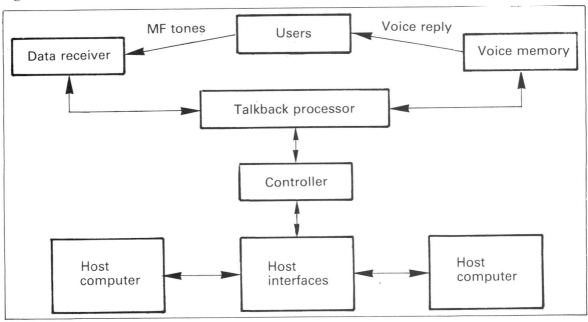

Figure 4.43 *The Talkback 'network'*

Autophone have installed a voice response system for Mobil oil, who have 1000 retail customers placing orders for fuel products on a 24-hour day basis. Access is through a Freefone number and the voice prompts customers to enter product requirements, delivery dates, etc. The IBM computer that backs up processing voice-echos the order detail to the customer so that errors can be spotted. It allows a customer to follow the progress of an order through all its stages with appropriate keyed requests and can also carry out 'real-time' credit-checking on orders entered. Plans are in hand to increase the range of customers that can be accomodated and to analyse traffic so that costs can be saved by a better understanding of order patterns, etc.

Other motor and motor accessory customers include Ford, General Motors, Fiat, Nissan, Toyota, Unipart and Talbot. Other customers include BT, British Aerospace, General Foods, Kays and Golden Wonder. Autophone have also installed a system for a Spanish bank who, as an inducement to their subscribers, supply a free tonepad.

One other application in particular is worth noting to demonstrate the power of the system. Brooke-Bond Oxo have an IBM 3032 mainframe at their Surrey main office which, amongst other applications, receives MF stock-ordering calls from over 250 salesmen, as well as further information from merchandisers and an indirect sales force. They realised fairly quickly that the original system implemented, which had salemen dictating orders to an answering machine, was too time-consuming and error-prone. A possible answer was the provision of conventional portable, data-entry terminals. This was rejected, mainly due to cost, but also because of inconvenience and the training needed.

They purchased a Talkback system and after a day's training, the sales reps 'went live'. Now, daily management information is available within 24 hours of bulk data being entered, and distribution loading and routing is much more streamlined. Weekly reports now reflect Monday–Friday rather than the Friday–Thursday that was the result of the earlier system of data entry. In addition, they save a lot of money on postal charges, telephone time and dictaphone usage.

In the very sensitive area of retail food sales, the increasing accuracy of sales figures and statistics, both for representatives and stores, allows the company to make fine marketing decisions based on up-to-date information.

Chapter Five

Text and document transmission

The techniques

Telex is the national/international network that allows for slow-speed transmission of text messages using a limited range of characters.

Electronic mail is a vague term used to describe the sending of text messages, conceptually or actually, as documents, over a transmission network large or small. At the lowest end, we could have a PC with several terminals, each able to send each other simple messages. At the other end, we have international 'message-board' systems which enable users all over the world to send and receive messages.

Teletex is a system for sending electronic mail at a reasonable speed, according to international standards and recommendations.

Facsimile (fax) transmission involves sending and electronic reproduction of a document through a network.

Data and text

Until relatively recently these two terms could be treated quite differently, not necessarily because of any real inherent difference (text is just formatted data), but because of the applications in which they apply.

Sending a message from one office to another, asking Blaise to give you details of some publication, asking your or someone else's computer to produce a list of debtors, etc., are similar in that the result can be thought of as a piece of text. Using a computer to process data usually implies a need to output the desired information in some convenient form, so that it 'informs'.

However, with telex, information access and electronic mail, it is not just a question of presenting the data in a reasonable form. We have the problem that to send a typical piece of text, we must think of it as having several different aspects. First of all, there are the actual characters that make up the words and the markers to show end-of-line, end-of-paragraph, end-of-message, etc. Then we have what could be called 'editing' signals which include commands controlling the actual format of words such as to change to bold, underline, italic, reverse video effect, video symbols, etc. An extension to this is the signals that allow changes of typeface and type size, which is becoming very important these days both in newspaper and book production. All of these are data to the transmission network itself, but the software at each end must know the difference. Protocols and standards govern the actual transmission of the characters, but additional standards are required, otherwise the receiver must have software to sort out the formatting.

For example, suppose we are sending a line that we want to appear like this:

Goods **must** be delivered our office, **Monday latest**. Fred.

To allow for its reproduction at the receiver's end (assuming his software knows what to do), the minimum that must be sent is something like:

STARTGoods [Bmust[N be delivered our office, [BMonday latest[N. Fred.END

The '[' character here would be detected as a signal that something has to change so it cannot be used within the text itself. The character after the [could indicate what the change is. In this hypothetical example, B means turn on bold face and N means turn it off. This is all very well for a simple example, but suppose we want a word in italics as well as bold, or force a new page, or a different typeface or super– or subscript. With all the different combinations, we will soon run out of letters after the [to indicate our choice. We may also want to indicate things like page numbering on or off, skip several lines, paragraph indent and so on.

It should not be too difficult for someone to set up a fairly simple set of rules and then write a piece of software that can scan a piece of text, insert all the control combinations and then, at the other end, recreate the original text and format. The problem is to produce a set of international standards that are independent of the transmission method, the software and the hardware, which are sufficiently comprehensive to cover the wide range of options but which are acceptable to everybody.

The name used for such a standard, as applied to electronic mail is *Office Document Architecture (ODA)* and both ISO and ECMA are working on draft proposals for ways of transmitting documentary text and facsimile, so that documents can be sent with the minimum of formatting once the text is received. The ISO recommendation is a draft proposal DP8613, and ECMA are working on ECMA-101.

We can also expect to see subsequent recommendations relating to specific text formats, such as spreadsheets, graphics (such as for sending bar chart and pie chart output from business graphics packages) and perhaps sound for voice response systems. Even further over the horizon may be the coverage of user directories.

A number of companies like Xerox, Hewlett-Packard and ICL have shown themselves committed to standards like these and some companies have defined their own subset of standards which they expect to be compatible with those of ISO and ECMA.

Telex is fairly simple in that apart from a set of letters (either upper case or lower case), digits, punctuation characters and a few special characters, nothing else is needed, so once a message has been typed in (or prepared on off-line paper tape or disk), it can be transmitted over the telex network.

Electronic mail is similar but much more attention may need to be paid to formatting. A further feature of electronic mail is the 'bulletin board' requirement where a user can scan a directory or private sub-directory to see if there are any messages, unlike telex where the document appears in the office of the receiver.

Telex

The telex network

Telex has been around a long time as a means of sending small quantities of text. This has been substantial development of the old telegraph system and, in the UK, was at one time part of the voice telephone system (PSTN). One of its main disadvantages is that the code used for representing characters only allows for about 55 different combinations.

From the early 1930s a separate *TELegraph EXchange* public teleprinter network was set up and in 1960 an automatic system was completed based on Strowger mechanical switches just like those used for voice. A link was also provided with the Telegram service (*TASS – Teleprinter Automatic Switching System*), but this was eventually abandoned.

The initial call was set up by telephone with a switch to teleprinter after connection was made. A user would phone through the details of the telex message to be sent. Later on, in order to reduce the time spent on-line, messages were prepared with an off-line paper tape punch and when ready, the tape could be fed into the reader attached to the teletype for transmission at six characters per second rather than at manual speeds. These facilities were provided by the old *ASR33*, this machine being in use for telex and as a slow computer terminal until not that many years ago. (Some telex machines still use paper tape while others use magnetic tape cartridge or floppy or hard disk or even computer memory itself).

The message-switching centre would then manually select the best route along which to send the message to its destination. A number might be batched together with others to an intermediate switching centre which could then split them up and forward them to the destination.

Once under computer control, the whole operation becomes quicker and more efficient although it requires much intermediate storage for messages sent and not yet received. During transmission, messages may be broken up into smaller parts and intermediate switches must be able to string them back together.

British Telecom are committed to a major upgrade to the telex network using the digital trunking that is going in. They have introduced SPC digital switching systems which provide a wide range of transmission speeds, from 110 to 9.6 kbps. Manufacturers such as GEC and Plessey supply the exchange and other equipment to BT and telex is now very sophisticated.

On the other hand, BT are conscious of the small, occasional user and, as was mentioned in an earlier chapter, the gateway *Interstream One* gives access to telex from a PSS terminal (and PSS access from a telex terminal).

Business and industry are the exclusive users of telex, there being almost no private subscribers and powerful facilities are demanded and supplied, both by BT and a whole host of manufacturers. In particular, in the period mid-1985 to date, everybody has become conscious of the fact that *inter-networking* is now possible with up-and-coming methodologies such as the use of gateways. This means that a quite typical user with, say, an IBM or other PC, wants telex from the same machine that is being used for word processing, spreadsheets, etc. If a telex has to be created, why not word process it? If the result of a calculation has to be sent abroad, use the spreadsheet and word process the figures into a message.

A further consideration is the use of electronic mail, viewdata and the fast up-coming teletex. Also with *InterStream One*, BT provide a gateway or inter-network link between telex and users on the PSS networks. It allows a PSS user to send messages to a telex terminal. Similarly, it acts as a gateway between a telex terminal and any PSS host computer. (Incidentally, the NUA for Interstream One is 2348 – *cf.* Switchstream which is 2342.)

The main advantage of telex is that it works to a universal standard and since there are about 100,000 users in the UK, with access to about 2 million users internationally, it seems that it will be around for quite a long time until a truly international standard electronic mail appears.

Telex terminals

As part of liberalisation, BT no longer have a monopoly on telex machines, although competitors still need BABT approval whether they supply the complete machine or as a PC attachment. If you are to become a telex user, you have a broad choice at present:

● Make use of a dedicated telex machine which has been designed to give all the facilities needed for an organisation to handle the range of telexes sent and received.

● Upgrade your personnel computer by attaching a 'telexbox' which will give you many of these facilities.

● Use an 'external' telex service such as can be provided from BT Gold or Mercury's Easylink (coming up soon under Electronic Mail).

The decision as to which you implement depends on:

● The volume of work to be processed, i.e. the number of telexes sent/received. Obviously the occasional telex can go through Gold, even though the unit cost of such an external facility may be high in comparison.

● The time-lag between you and the people you telex to/from, i.e. do you want to set up a string of telexes in advance and send them off at pre-determined times. Similarly do you want automatic receive outside office hours.

● The sophistication of the service required, e.g. having a word processor available to format telexes, being able to include standard paragraphs, retry if busy and send multiple copies ('multi-addressing' to different receivers).

- The different types of people who will make use of the machine. A PC keyboard is not usually designed for the professional typist, but on the other hand, micros these days can have large disks and memories which mean that the operator need not worry about changing disks or running out of memory work space.

- The other facilities wanted from your 'telex machine' such as database access or videotex, or integrated packages, which will require the use of a PC.

- Cost. Several suppliers offer kits that extend a PC for between £1000–1500 and if you already have a PC, this could be the cheapest route.

The range of facilities wanted

- Ease of use, in particular a menu for selecting facilities. Figure 5.1 shows the main menu displayed by the Hasler Protelex unit running with an IBM PC. The kit operates under the CP/M operating system used by many micros. Using the special function keys, you can select:

- *Automatic dialling and answerback checkout* As with the intelligent telephones, it is very convenient to have short-code or dedicated key dialling in order to make immediate connection and handle the checking of the answerback code at the beginning and end of transmission. Another important feature is automatic redial if the caller is busy.

- *Short-code (abbreviated) dialling* As with telephones, it is efficient to be able to code a telex number and answerback by a short group of characters. Even more useful is a kind of 'supergrouping' whereby a short code could access many numbers classified by geographical area, type of customer, etc. The main advantages of short codes appear with international calls where the number may be quite long (and very easy to get wrong – if the answerback code fails to tie in with the telex number, you are still charged for the connection time to find this out).

- *Multi-addressing* Particularly with something like telex mailshots, it may be useful to send the same telex to many different companies.

  Conversational ('wild' telex) Sometimes you may want to send a telex immediately or one that is special in some way, such as not in the short code libarary. Alternatively, an actual two-way telex conversation may be wanted.

- *Automatic message reception* Messages may be accepted for display on the monitor printer, diversion to a separate, slave printer or storage on disk for display when convenient.

- *Simultaneous create/send/receive* Being able to set up a string of telexes for later dispatch, while earlier messages are being sent and incoming messages being handled. This can be extended so that a PC with attachment can send and receive telexes overlapped with other electronic office functions or even applications running such as payroll.

- *Access to input and output queues and the usage log* Having set up a stream of outgoing telex messages, it is very convenient to be able to scan through a display of the queue to see what you have done and, at the same time, interrogate the output queue to see what has arrived so that you can direct it to the relevant printer. Similarly, it may be necessary to have access to a log which will show unauthorised use and give some measure of proof of sending a telex.

Function key	Function
1	Prepare text
2	Send a background telex
3	Send a conversational telex
4	Monitor incoming telexes
5	Respond to an incoming telex
6	Display telexes
7	Delete telexes
8	Display log
9	Display queues
10	Miscellaneous

Figure 5.1 *The Hasler Protelex menu*

- *Batching* After looking at the outgoing queue, you might spot that there are several to the same company and it will save money to be able to link them together as a batch for transmission.

- *Reverse batching* A fairly new trick where your caller can request a batch of telexes to be sent to him.

- *Priority assignment* It may be useful to be able to give a rating to a recently entered telex so that it goes before earlier ones. A slightly different aspect is the assigning of output priority over input. This could be important if you only have one line and are in the middle of sending an expensive telex which you do not want interrupted. Some systems provide several levels, including an absolute over-ride that breaks incoming or outgoing transmission to do what you want.

- *Message word processing* The message that goes is just text and a good word processsor can drastically reduce the time to prepare a message. Even more important could be the creation of part of the telex from data generated by an integrated package. There should be no problem in using a word processor to select spreadsheet or database query output for incorporation in a telex.

- *Library access (pro-forma telex)* Automatic access to a library of standard telexes or perhaps paragraphs or common phrases will again reduce preparation time.

- *Date and time* The machine should keep the date and time-of-day automatically and tag them to outgoing telexes. A facility is needed to reset them when necessary. Some systems have a neat way of allowing you to enter it into the body of the message automatically. This could be nice when you are calling up standard text from a library: *'Your order received by us plonk will be dealt with …'.*; *'plonk'* is replaced by the current date.

Stand-alone (dedicated telex) v. telebox

There is little doubt that many companies will continue to use dedicated machines for quite some time to come. If the requirement is for sending many fairly short messages to many customers, professional operators will be employed, who will want the telex keyboard they have been used to and because of their experience, they will be happy not to have many of the facilities above because they will seem like gimmicks.

Stand-alone telex

A stand-alone telex machine might be expensive, although 3M have recently introduced their Whisper telex machine which is really quite tiny compared to other telex machines or even PCs. It does not have a VDU, but is fitted with a silent, thermal printer and offers a number of the features you might want. It is obviously limited in application in that it does not have disk back-up, but it does have a useful little memory and there are a number of special functions available from the keyboard.Whisper is available in send/receive or receive only models.

A neat little stand-alone machine comes from Olivetti called the *TE530* which is about the same size as an electronic typewriter. It has a 16-line digital display window and a separate key pad with 10 special functions, such as directory-print, automatic print-out, edit and search. Disk and paper-tape are available as extras.

British Telecom offer a range of telex machines including *Sable* which is basically a small, receive-only teleprinter which can be upgraded to send with an add-on keyboard. It is microprocessor-controlled and has facilities like short code dialling. It is fitted with a matrix printer which produces incoming text in an italic form to make it stand out from the outgoing. Normal upper and lower case is available off-line, such as when using the machine as a typewriter. BT claim that an operator can pick up usage of the machine in an hour or so. Sable sits on the desk or on a specially designed plinth.

The *Cheetah* (made by STC), is a much hairier animal intended as a sophisticated single-user telex machine, but there is available a communications interface which enables connection to the facilities via some PCs, electronic typewriters, etc. The system has a large-character integrated or separate VDU and a keyboard fitted with an

Figure 5.2   *The 3M Whisper telex*

extra pad that allows word processing like editing and correction facilities while telexes are being sent. Optional floppy disk back-up is available as is paper tape if wanted. The Cheetah offers quite a few of the desirable facilities including simltaneous receive and prepare.

Figure 5.3   *The Olivetti TE530*

A more powerful machine is the *Puma* which is something like an extended Cheetah with the communications interface. The 'Mailbox' facility allows any terminal within the building to access by 'posting' to/from the Puma and this could be extended to a distant remote terminal provided the communications hardware is compatible.

The machine does not have a VDU as standard, but is equipped with a 40-character LCD buffer. This means that when preparing text, a line goes into the buffer first and can be corrected there if necessary. It can then be released into the system. Paper tape is available for off-line message preparation.

A much smaller-scale machine is available from WordNet called *TelexMan 20*. This allows for remote telex preparation from various micros, typewriters or dedicated word processors. It has user-programmeable keys for useful phrases and

Figure 5.4   *BT Merlin Sable telex terminal*

Figure 5.5   *BT Merlin Cheetah Teleprinter*

Figure 5.6   *BT Merlin Puma telex terminal*

Equipment Telex 2000 and the telex units supplied by Hasler.

Companies such as Transtel offer a range of stand-alone telex units from their *ComWriter* up to a sophisticated disk-based system called *ComMaster*. The ComMaster can be supplied with an on-board, English/Arabic translator. Figure 5.8 shows the machine in use and Figure 5.9 is a close up of the two windows on the screen.

can cope with pro-forma layouts and standard paragraphs.

Telex using a PC

The very small user can use a simple PC such as a Commodore or BBC and hook into telex via Gold or Prestel. Alternatively, there is a service, which has been supplied for many years by the company British Monomarks, that offers a letter mailbox facility, now extended to electronic mail with telex access.

The small user who has an IBM PC, has quite a lot of choice, such as the Data and Control

Figure 5.7   *Transtel ComWriter III with matrix printer and keyboard with autodial keys*

Dataline Systems sell a telex package called *Streamline* for use with IBM-compatible PCs. The sofware called *pcTortoise* is on a printed circuit board called *pcStreamline* and plugs into the PC. It comes with its own power supply and printer and can receive telexes independently of the PC. In the event of error, a buzzer sounds for the user and incoming calls get an engaged signal. It would appear to have a large number of the desirable facilities. If the small user happens to have an ICL OPD, he can hook in a *Telexbox-3*, so that the features of the two can be combined.

Braid Systems Ltd offer their *Telex Manager* and *Mail Manager* software/modem packages which

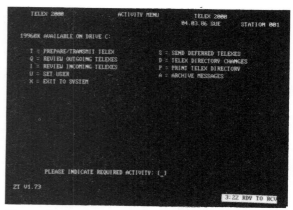

Figure 5.10 *Telex menu from the DCE Telexbox-3*

run with several micros under several operating systems.

Some companies, such as Integrated Business Communications provide very powerful sets of software and hardware which can convert a PC into a telex machine. With their kit, you have most of the desirable facilities and a choice of the number of terminals and lines you want to use. (Incidentally, this company can offer a similar package that turns your character terminal PC into an X25 packet terminal.)

STC have developed their *Textel* range of telex processors. They come with a variety of facilities

Figure 5.8 *Transtel ComMaster telex terminal*

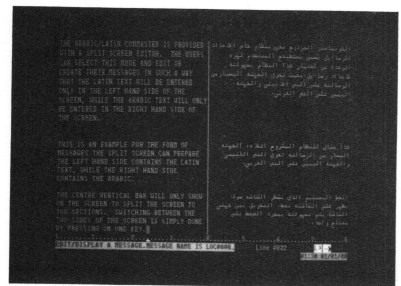

Figure 5.9 *Close-up of English/Arabic translation*

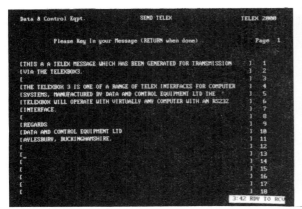

Figure 5.11 *A telex in preparation*

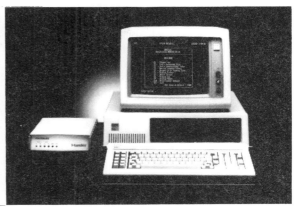

Figure 5.13 *The Hasler telex with an IBM PC*

Figure 5.12 *An outgoing telex queue for examination*

Figure 5.14 *A DCE Telexbox-3 with an ICL OPD*

including disk back-up, extra printer ports, linkage for other terminals, a purpose-built word processor and the general ability to upgrade the system as your requirements increase.

The next step might be to use a message switch such as supplied by all of the above companies (among others). These all allow for several message-preparation terminals. The largest machine, offered by Fenwood Designs and called the *Stofor*, can have up to 20 ports and, in addition, can have a 10Mb hard disk holding a 2000-address library and can handle computer-derived input at up to 2400 bps.

Telex switches include advanced facilities so that PCs and even electronic typewriters can act as terminals and input/output can be multiplexed onto one line.

Ferranti produce *Telex Manager*, a very clever piece of hardware which as well as providing most of the 'desirable facilities' makes maximum use of telex lines from a range of terminals and, in addition, has a 'partitioning' facility which means that several companies in the same building could make independent use of the same Telex Manager.

Electronic mail

This is a very vague term that can cover anything from telex, through PCs talking to each other across an office, right up to services like BT Gold and Easylink. In addition, it is intimately linked with the term 'electronic office'.

The concept evolved as a method of sending text

files under computer control so that the user need not actually have to go to the trouble of creating a computer file as such and neither he nor the receiver need any computer knowledge. As we have seen in earlier chapters, data has been transmitted commercially from computer to computer and computer to terminal for over 20 years. But using such facilities for text, i.e. documents, is much more recent, largely because of the costs of transmission but also because of the lack of adequate low-cost software. Don't forget by the way, that 'hard copy', i.e. printed copy, of electronic messages is usually needed, both for some proof of delivery/receipt and for future reference; an electronic mail service should provide this when required.

Electronic mail facilities

Let us first review that range of services we might want from electronic mail/electronic office:

- *Word processor* Easy production of letters/documents – for the experienced typist secretary in order to increase efficiency and convenience dramatically and for the 'casual' user, to make use of the rub-out key and the various other functions that help you to correct mistakes, so that the usual scrappy note or letter becomes fairly presentable.

- *Spreadsheet* A sophisticated, screen-based calculator, especially useful for handling tables of data.

  Database A simplified version of what a mainframe computer can provide in storing, manipulating and retrieving/presenting data.

- *Graphics* A program that can take input from database or spreadsheet and present it as graphs and diagrams, sometimes with statistics calculations such as means and standard deviations. Originally designed for VDU display, people these days want printed output and some terminals are equipped with a 'print key' which is a signal to the system to dump the contents of the screen onto a printer. Even more effective is to transfer the output to a colour printer or high-resolution graph plotter (which can now be purchased for not that much more than a dot-matrix printer).

- *Diary* Exactly what it says; you or your secretary enter bookings and appointments and, on request, the program gives you a list or warns of a double booking. Even more useful with multi-user systems, with the ability to interrogate other people's bookings and get the program to find out when other people are free.

- *File 'importing'* A big feature of integrated packages, importing means being able to transfer files from database to spreadsheet to word processor, etc.

- *Viewdata/Videotex* This gives terminal access to useful data in a fairly standard form, either from your own company or from an external provider.

- *Bulletin board* As it says, an electronic message board which you can scan for 'messages' left for you and to which you can post messages for others.

- *Teletex* A relatively new service for transmission of correspondence.

- *Telex* As we have seen, some telex access is needed by many firms.

- *Facsimile* Transmission of whole documents by scanning and conversion to electronic impulses (dealt with later on in this chapter).

Note that *Teletext* is not included in this list. This includes the UK services *Oracle* and *Ceefax* which are primarily intended for private receive-only users with special TV sets. We will say a little more about them in the viewdata chapter.

Although there is quite a lot of electronic office software around, we can think of most of it as being internally oriented, i.e. to users in the same building. As with telex, many user installations allow several terminal users to send telex and electronic mail messages to a central transmit/receive facility, usually an operator, rather than give everybody a telephone line, a telex line and all the necessary hardware/software.

But, if BT have their way, there will be an increasing trend to *Teletex*, an integrated external service which provides a standard way of sending/receiving most forms of correspondance using the new transmission technologies and bringing together many of the electronic office requirements.

Before looking at teletex, we must deal with the three main 'external electronic mail' facilities.

One-to-One, Easylink and Gold

One-to-One comes from a US company operating in London as One-to-One at: Scorpio House, 102 Sydney Street, Chelsea, London SW3 6NL.

Easylink is from Cable & Wireless Easylink Ltd, Smale House, 114 Great Suffolk Street, London SE1 0SG.

Gold is from BT Telecom Gold Ltd, 60–68 Thomas Street, London SE1 3QU

They offer similar facilities with different customer charging mechanisms. No attempt will be made to compare them on a cost basis. One-to-One and Easylink offer the required communications for most makes of micro on the market.

One-to-One

- Direct-dial or One-to-One national PSS network (local rates).

- Log and queue inspection. Message access by password only.

  Telex.

- Message broadcasting and mailing list.

- Priority letter – copy-typing and first-class letter-posting of your messages to companies not subscribing to One-to-One.

- Courier service for priority letters.

- Radio-paging.

- To come – the company expects to provide access to many of the videotex services and home shopping/banking in the near future. They will advise on the suitability of your micro

Figure 5.15 *One-to-One in use from a mobile terminal using a cellular radio link*

Figure 5.16 *One-to-One electronic mail*

or electronic typewriter and modem (or accoustic coupler). A special autodial/autoanswer modem can be supplied.

The typewriter company, Brother, have recently put out a very neat micro microcomputer, the EP-44 – a tiny portable with mains unit, accoustic coupler and free subscription to One-to-One. The machine also has an integral 16 chps printer. (Several other hardware/software suppliers offer subscriptions to One-to-one and Easylink as part of their package.)

Easylink Similar facilities to One-to-One. Specials include:

- *Transatlantic services* including Easylink-to-Easylink subscriber in the USA cheaper than international telex and 'Mail-gram', very much like One-to-One's priority letter, but to the US for companies without telex or Easylink. You give the message and the address/zip code for maximum 24-hour delivery.

- *Dial-a-gram* – if your terminal is down or you are out of the office, you can dictate a message to Easylink operators who pass it on in the normal way.

- *Translation service* – messages or telexes into most European languages

Gold is slightly different in that all the 'intelligence' is at the other end. All you need to access its services is an input/output device (such as a VDU/keyboard) and an appropriate modem. The communications software is supplied within the system itself. Access is via PSTN and PSS and facilities include:

- *Message storage* Figure 5.17 shows the login dialogue to Gold and the user requesting a list of messages held (in this case, just one). The display was produced on a ELF supplied by Easydata Ltd.

- *Telex*.

- *Radiopaging*.

- *Business services* Daily bulletin of corporate developments, air travel and hotel informa-

tion, translation, typesetting, word processing and graphics.

- *On-line 'Chat'* – interactive user messages.

- *Miscellaneous* – Diary/calendar scheduling, noticeboard, phone message reminder, forms design and processing and a database management system.

As you can see, Gold tries to offer a whole range of electronic office services as part of a corporate offering and appears to be rather more comprehensive than the other two. However, it remains to be seen whether users will go for the wide range of applications that are available from and within Gold or the more limited choice from the others but with the flexibility of selecting your own software and data sources. In other words, if you want to use WordStar, Lotus 1–2–3, Open-Access, Xchange, etc., to file and process your data, link in public/private videotex and send extracts as messages or telexes, you will possibly be happier with the other two. If you are a small organisation with limited capital resources and you do not already have an investment in fancy software and hardware, Gold may be the best choice.

To give you some idea of this, several companies provide software to enable the BBC microcomputer, which is still one of the cheapest on the market, to be used as a Gold terminal.

At least two other companies offer similar services including GEISCO, with their *QUICK-COMM* and Istel with *Comet*.

Finally, a couple of further points. For the large user, Ferranti provide their *VM600*, a large voice-messaging switch for use with PABXs with up to 1000 users. The system is for voice messaging, not electronic mail, but it seems appropriate to mention it here because it has the ability to store messages in what Ferranti call 'Mailboxes', which can be partitioned for storage in 3–36 hour boxes depending on the load on the system.

The software company ADR supplies a product called *eMAIL* which can be used within an existing internal network to provide electronic mail facilities, with restricted access messages, accounting, reminders and so on. A central monitor

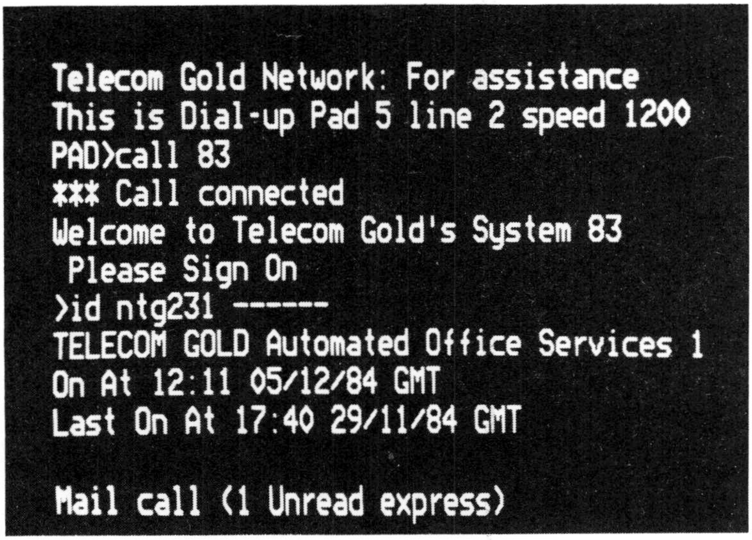

```
Telecom Gold Network: For assistance
This is Dial-up Pad 5 line 2 speed 1200
PAD>call 83
*** Call connected
Welcome to Telecom Gold's System 83
 Please Sign On
>id ntg231 ------
TELECOM GOLD Automated Office Services 1
On At 12:11 05/12/84 GMT
Last On At 17:40 29/11/84 GMT


Mail call (1 Unread express)
```

Figure 5.17 *Initial Gold dialogue and message display*

controls the receipt and distribution of messages to users within the organisation.

Another point to note is that Prestel has a message/mailbox facility as part of its general services.

Finally, it is the author's personal view that before very long software companies will realise that large-scale electronic mail provides a superb vehicle for contract data processing and it seems a certainty that some kind of basic ledger accounts processing (invoicing, budgeting, etc.) will soon be made available, as a logical extension to home banking and home shopping.

BT and Douglas Information Systems have recently formed a company called Edinet to promote *Electronic Data Interchange (EDI)*. The aim is the electronic transfer of documents like statements and orders between computers. One wonders if this might eventually encompass a similar facility for electronic mail users as well.

Teletex

Teletex is a high-speed (low-cost) text communication service which, since 1985, has been sponsored by the Department of Trade and Industry and heavily pushed by BT in the UK as well as by other European companies, in particular Seimans in Germany. The idea is to provide a network based on the PSTN and PSS which gives easy, gateway access to the telex network as well as foreign teletex networks. BT are already making noises that teletex will drop quite naturally into the currently nascent ISDN and they refer to the term *teleport* as the customer terminal access point through which teletex IDA can be obtained to all the proposed BT and other digital services. They talk of a 'premium' sector for the larger customer with leased lines and an 'economy' sector for the smaller user.

It is very likely to succeed mightily, in spite of a rather slow start because of what appear to be very attractive transmission speeds (up to 30 times faster than the 80 words per minute of telex) and correspondingly lower charges. In addition, a full 309-character set is available, i.e. anything usually available from a normal typewriter keyboard including subscripts and superscripts, removing the upper– or lower-case restriction of telex and allowing for 'foreign' characters such as German and Scandinavian. Another important factor is the fact that the protocols associated with teletex involve a considerable amount of error checking (error-free documents received?).

Although many countries, including the USA, the Scandinavian group, Canada and Australia, have their own teletex services, they tend (except for Norway) to be based on circuit-switched data

Figure 5.18 *MerlinTex adaptor with MT4000 microcomputer*

networks and until these countries introduce the relevant inter-networking equipment (in accordance with CCITT recommendations X72 and X75), true international teletex may have to wait for a while to be really successful. (The proposed undersea optical-fibre network, mentioned in Chapter 2, could help in this – agreement on tariffs will hopefully not take as long as agreement on standards and protocols.)

We should put teletex into perspective by looking at what it can offer compared to using a mix of other facilities.

Teletex as applied by BT operates on the PSTN and PSS using gateways.

You will recall Interstream One, mentioned above, as the PSS-telex gateway. BT now offer

Interstream Two which can link a PSS teletex terminal to a PSTN teletex terminal, and *Interstream Three* which links telex and PSTN or PSS teletex. With this, an optional positive acknowledgement of teletex-to-telex calls can be obtained.

PSTN access will be along dedicated lines which, being a simpler form of access, means that terminals are simpler and should be cheaper and more reliable. Also, they will not compete with voice transmission. Transmission rates can be at up to 2400-bps, although 1200-bps terminals will be compatible. PSS terminals can operate at 2.4, 9.6 or 48 Kbps, but the system can be matched with the receiving terminal.

Most PCs can be used with a suitable adaptor. BT can supply *MerlinTex* which they offer in conjunction with their Merlin word processor, but it will link most common micros into the system and even electronic typewriters. Fig. 5.18 shows the adaptor in use with the BT Merlin M4000 microcomputer.

Ferranti also supply a teletex adaptor which allows autocall and autoanswer and provides speeds from 50 to 9600 bps. It is intended for micros and mainframes to cover the larger company that already operates an internal, corporate electronic mail service. It also has a printer port if the full teletex character set (CCITT recommendation T61) is beyond the user terminal.

For the smaller user, Ferranti, like BT Merlin, also supply a teletex IBM-based PC which comes with database, spreadsheet and a range of printers.

To get into teletex via an adaptor, the 'host' operator prepares the text of a message, attaches a destination and passes a copy to the adaptor. This inserts the receiving identity, the calling identity and transfers the completed message to a store. It then calls the remote destination and carries out an identification 'handshake'. If successful, the text is sent and the details recorded on a log.

Incoming messages are acknowledged and stored with a warning that a message wants to be printed.

Philips Business Systems market an automated business system kit which handles telex and teletex. For example, it is possible to broadcast a message to a number of potential teletex receivers; any that cannot receive teletex are automatically sent a telex instead.

The following list (courtesy of BT) show companies that supply terminals that are verified for use on for the BT teletex service:

- Ericsson *Eritex 10*: typewriter with 40-character display.

- Ferranti *Teletex Adaptor*: add-on for PCs, etc.

- GEC *Centex*: multi-terminal support device.

- Olivetti *TLM 601 Adaptor*: Olivetti machine interface, also for IBM PCs.

- Philips *PS5020*: word processor with disk support and printer.

- BT Merlin *M4000*: word processor, floppy or hard disk.

- Siemens *T4200*: electronic typewriter with screen and floppy disk

- STC *TX4000*: teletex adaptor.

- Symicron *Symex*: teletex adaptor.

The following companies expect to gain verification of their equipment: AES Data (UK) Ltd, Mitel Telecom Ltd, Olympia, Plessey Office Systems Ltd, Transtel, Triumph Adler, Rank Xerox.

MHS

Before dealing with fax we should mention a very recent electronic mail facility announced by BT. They call it *MHS (Message Handling Service)* and it is hoped that it will provide a service transparent to customers whereby messages can be sent from caller to receiver instead of having to make use of a mailbox which needs interrogating.

It is hoped that the service will link PSS, teletex and telex users as well as corporate private networks with an external messaging capability. The general idea is to implement an integrated service rather than allow gateway access from one to another following the recent X.400 CCITT recommendation, which sets out the protocols that will be involved. It remains to be seen how the introduction of MHS will affect the future of teletex and telex.

Facsimile transmission (fax)

Facsimile transmission is very different from other methods of 'document' transmission in that, instead of sending the characters that make up the text, the whole document is scanned and the contrast between light and dark is coded for transmission. In other words, as the scanner passes across the document, darkness above a certain level is given say a '1' significance and below, the opposite. The 1/0 is registered by frequency or amplitude modulation.

The electronic image is then sent as a series of analogue signals like speech, or in the more sophisticated machines, as a series of binary digits to be reproduced at the other end where say, a 1 gives a dot and 0, no dot.

Line speeds employed are 2400–9600 bps although some systems make use of 'fast' protocols which maximise the actual transmission rate when sending between two identical models. The image resolution obtained varies depending on the model and is usually expressed as the number of dots to the inch. The technology has been around for quite a few years but equipment was very expensive and limited in use to large organisations like Interpol, US police forces, etc.

More recent times have seen fax growing, especially in Japan, the largest highly developed country without a Roman alphabet. The Arab countries have a similar problem but much easier to solve since, even though written Arabic does not use Roman letters, the script does consist of characters that can be coded in, say, ASCII, and displayed on a VDU or printer. Written Japanese, like Chinese, is made up from a large number of pictograms which are too complex to be coded and sending an actual facsimile (copy) of the original document is an obvious solution.

Particularly since 1984, most suppliers work to CCITT standards, compatibility is becoming less of a problem and fax is really becoming more widespread. (It is estimated that over 50,000 machines were installed by the end of the first quarter of 1986.) It is a little difficult to see why though, in spite of what the manufacturers claim. For someone who wants to send an exact copy of a drawing or diagram such as a cartoon, fashion drawing, piece of engineering or of text where the exact format is important, fax is marvellous.

To give an excellent example, this book has been set using a very powerful word processor. The text after formatting on disk is printed out in actual page format from the word processor in exactly the form you are reading except of course for the addition of diagrams and photographs. These are then completed by the addition of the diagrams and whole pages are then faxed to the printers.

Some of the more recent newspapers also make heavy use of fax.

Another application is when a customer wants a fast copy of an invoice or contract, especially since it is much more difficult to corrupt fax criminally during transmission compared to sending characters by electronic mail, etc. (However, the 'Postal Rules' which are the most recent legal framework relating to contracts sent by electronic means, only apply to telex, so it is difficult to forecast the result of legal action based on delivery/non-delivery of fax or teletex.)

But as a means of sending just text, such as a copy of a letter or a simple message, it would seem to be a bit of a luxury although it is obviously convenient to have a machine into which you can place a document just like a photocopier and then literally press a button to send it off.

Regarding charging, although telex does not have cheap rates, the costs depend on the time actually used, rather than to the nearest minute with fax (which is using the telephone lines). Another aspect is that with fax, charging depends on the size of the document rather than the number of characters printed on it. And even though reduction and 'white paper ignore' is available on some machines, it is very tempting to put in even a short message into a fax machine and this must be controlled to keep costs down. Nevertheless, let us look at what is available and then try to predict what might happen to fax.

Fax 'groups'

Since 1968, the CCITT has laid down four levels of standard relating to fax which can be summarised thus:

Group 1 Slow scan (A4 sheet in 4 to 6 minutes), with a resolution of 100 dots per inch. (The scanning process makes use of frequency modulation as part of the light/dark coding). Group 1 machines are dying fast in favour of the other groups.

Group 2 A4 sheet in about 3 minutes with a similar resolution and using amplitude modulation for coding which gives an image with a better contrast than Group 1.

Group 3 About twice as fast as Group 2 and with about twice the resolution. The scanning generates digital signals which will obviously fit in nicely when fully digital networks are available.

Group 4 This is based on digital technology and is still being worked on, but the aim seems to be for speeds about 15–20 times faster than Group 3, making use of a laser printer for the actual output. The idea behind Group 4 is to utilise packet-switching facilities.

Informed sources think that we are about five years away from widespread Group 4 usage, with estimates of over £11,000 for basic equipment at present. (It is interesting to note that the company Fax-line Servicing have recently launched their *Teqcom 1830* converter for use in private networks which may be utilising packet-switching. The unit has two inputs: for fax and computer terminal. The point is that the fax machine can be addressed by the terminal and this allows you to attach a fax image to a piece of text that is being sent by electronic mail. To make use of a dedicated switch for fax to use packeting would make the unit cost of sending fax extremely high, but the Teqcom can drastically reduce this.)

Visual reproduction methods

The early machines generated visual output with a stylus which chopped out the dots from the impregnated surface of the paper. More recently, electrostatic printing, similar to photocopying, has been employed using the same kind of toner. Many machines use thermal printing, where the paper is heat-sensitive and a dot is produced by a local heating element. Occasionally, ink-jet printing is used (much like the ink-jet matrix printers used as output devices for computers).

The most recent trend has been the introduction of laser printing which gives very much clearer image production.

The range of facilities wanted

What is needed in a fax machine depends on the user and his requirements, but like telex, we can look at the facilities that you might want:

Immediate photocopy It is useful to have an off-line photocopier as well as a fax machine, but more important, with a bad original, you will get an idea of the quality that will be received at the other end.

Flat paper feed If you need to protect your original from being bent, it is a good idea to use a flat-bed feed machine which feeds originals horizontally. A further advantage is that you can load a magazine or book, in some cases up to about a quarter-inch thick. This can save your having to transmit a photocopy of the original.

Automatic document/paper feed For the larger user it is important to be able to load a batch of documents, preferably allowing for different sizes, and similarly to have automatic feed of blank paper for the final images.

Variable original size/reduction-enlargement of documents These days many people make use of standard DIN paper sizes, especially A4 and A3, and some machines will accept a mix of documents for transmission. It can be convenient and cheap to reduce an original before transmission especially if the receiver cannot take a large document. When sending a small document, it may be appreciated by the recipient if you can send him an enlarged copy. Some machines can automatically select the optimum size for transmission and others allow the feeding of documents up to 10 feet long.

Group 3/2 compatibility Although Group 3 machines are becoming more common, to be of maximum use, they need to be able to talk to machines of Group 2.

Automatic send with short code or coded key dialling For often-used numbers it is a big selling point to have abbreviated dialling from directory. For example, the Canon *FAX-520* allows 24 pre-set keys to be assigned with 50 further numbers by a short code consisting of * followed by a two-digit number.

Automatic redial If a number is engaged, many machines will retry several times.

Talkback/terminal identification This is important to ensure that your document has been received and for security. Like telex you can make

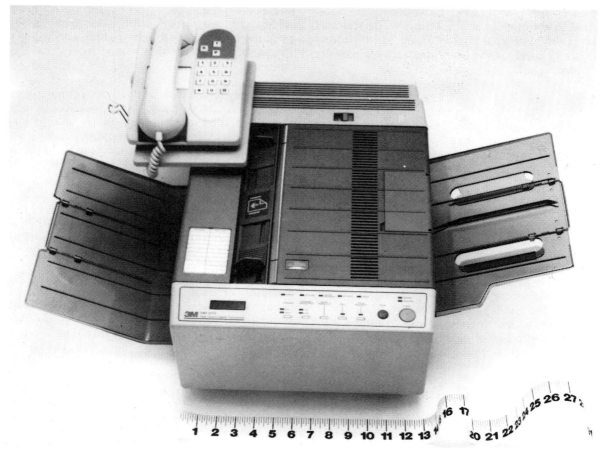

Figure 5.19 *The EMT 9715 fax transceiver from 3M*

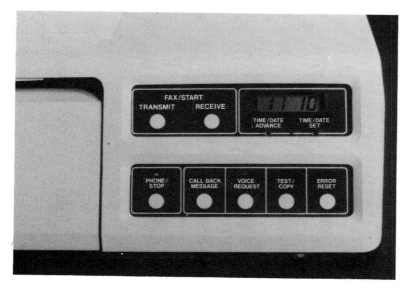

Figure 5.20 *Touch panel from
the 3M EMT 9146*

use of an identification/password check-out routine to ensure that the receiver is the one wanted and to prevent unauthorised use of your machine either for send or receive. For received documents, the better machines put a heading on the top showing date, time, sender identity/code and a page number.

Error indication and auto-disconnect.

Polling This means being able to contact a remote sender in order to request a document to be sent if ready to go. Reverse polling allows your machine to poll your receiver as soon as you have sent your document so that he can return if he wishes.

LCD display To see what is going on, especially when dialling out, a number of machines are fitted with a liquid-crystal display panel.

Automatic selection of transmission speed Assuming that the modem can handle it, some machines select the highest speed to ensure quality, reducing it until the quality is obtained.

Talk attention/voice request As a document goes off, you may want to signal the receiver to pick up his phone so that you can have a voice conversation.

Deferred transmit/document memory This can be very useful and, like deferred telex, allows you to use cheap rates or send across time zones. The Canon *FAX-710*, a development of the 520, can hold seven pages of A4 in a 1 Mb memory which can be expanded to hold up to 28 pages.

Unattended receive With long paper rolls (up to 100 metres in some cases), and on-board guillotine, there is often no need to have an attendant to receive images. About 300 A4 pages can be produced from a long roll, with date/time attached.

Copy quality control/fine-superfine mode To save transmission time, if the required image does not need to be too clear, you can often select a low or high scan. The low scan in effect reduces the number of dots to be printed and hence the number to be sent.

Figure 5.21 *MerlinFax transceiver*

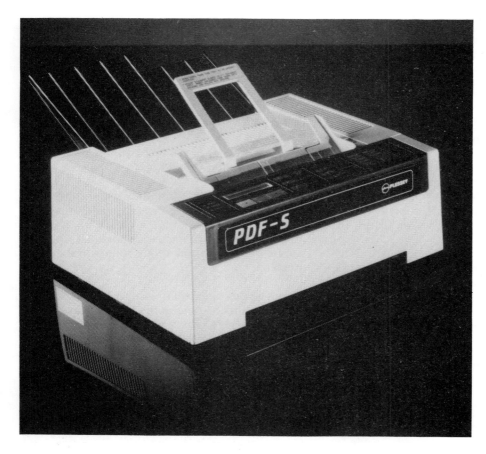

Figure 5.22
The Plessey PDF-5

Fast mode This is based on compression/decompression techiques, where 'white space' and repeated characters are coded so that they are not actually sent as they appear. This can reduce transmission times by 30–50% and is often offered between similar models from one supplier.

Background control Particularly when sending poor photocopies or documents with fuzzy colour background, it is useful to have a facility to increase/reduce the contrast sensitivity.

Talk reservation/line hold After the document has been sent, it is useful for the line to be left open so that you can talk to the receiver.

Activity journal/log Useful for cost and security control, a number of machines produce a printed message with each transmission. The *MerlinFax* machine, for example, has this facility and in addition, produces a transaction log print-out on request or after every 32.

External input port for character-to-fax A feature that is starting to appear with some sophisticated offerings, this allows text to be input from wordprocessor or other computer terminal directly into fax. As a very advanced extension to this, mention must be made of the Rank-Xerox *Netmaster*. This is something like the message switch used for telex and, as well as being compatibile with Groups 1–3, it enables a link to fax from word processors, micros or even mainframe computers as well as their own *495-1* and *295* fax machines. The Netmaster has hard or floppy disk back-up and it is claimed that the system can hold for transmission up to 100 A4 pages, with a 999-destination library.

Rank Xerox also produce a neat, portable Group 2/3 machine that weighs about 22 lb. It comes with a 24-number memory telephone and is said to handle about two pages per minute.

Canon offer several machines. The *FAX-520* has

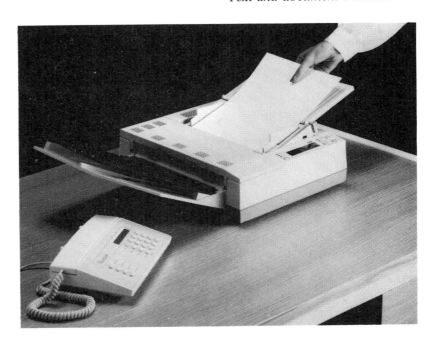

Figure 5.23 *Xerox 7010*

Figure 5.24
Canon FAX-520

its control panel at the front and incorporates an LCD display of date, time and other explanatory messages, a dialling pad and 26 autodial preset keys. It can be linked to other fax machines and poll round them collecting fax from them on a pre-timed basis. It will send B4 or A4 and will reduce if the receiving machine cannot handle B4.

The *FAX-220* is a neat little desktop machine and the *FAX-720* is for high-speed application. It has 20 preset keys but can hold a further 100 directory numbers. If a called number is engaged, the machine will try again after 1 minute and then after 2 minutes. A report is produced if all three calls fail.

Scanning/digitising to computer files

A trend that could affect fax very seriously may be sparked off by the advantages to be offered from digitisation of characters rather than whole documents.

The Xerox 295, like other machines, has disk back-up and, in addition, has very high-definition read-write and special-purpose circuitry which allows it to send output to a computer file for later printing. Used within an internal network, this might eventually be extended so that a user could have access to a library of fax images and with a suitable 'word processor' changes could be made and then the document sent from disk, via a fax machine.

In fact, Abaton Technology Corporation market a very high resolution character scanner (300 × 300 dots per square inch) that generates digital output for disk storage. The system is compatible with IBM PC and the Apple *Macintosh Plus* computer. The Macintosh has a program called Macpaint that allows diagrams and drawings to be produced with a similar level of ease as a conventional word procesor gives for text. In addition, Apple supply a laser printer which has a resolution equivalent to the Abaton digitiser.

If you already are making use of a micro of this kind, whether part of a local-area or wide-area network, the extra cost of the digitiser gives you the equivalent of a fax machine as well. Obviously for this high level of resolution, the number of bits that must be sent is very high, but this will become less of a problem as ISDN takes over. The overall

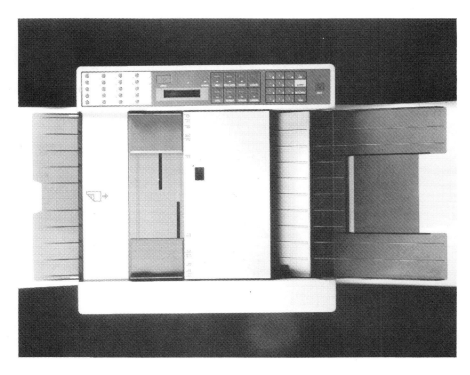

Figure 5.25
Operation panel of Canon FAX-720

costs of this seem to be comparatively high at present, but for the larger company with a requirement to fax documents to many different branches and hence many fax machines, this could, if properly controlled, become a very convenient and labour-saving facility. Such a company will already have the transmission network and can probably justify several laser printers anyway, so the sending of digitised documents could be a viable alternative to fax.

The major constraint for the near future will probably be the lack of suitable standards, particularly relating to different printers, so communication with other, external systems might be a problem until some degree of standardisation evolves.

Even more recent developments have concerned scanning to computer file from books and microfilm as well as documents and software is being developed to file and address 'documents' and to edit text and graphics. Optronics Ltd offer a system that involves a microfilm digitiser, a flat-bed scanner and a book scanner. Data is entered from the source document and compressed. It can be computer-clarified (using statistical optimisation techniques) and stored on magnetic tape or video disk, for later transmission or sent as data via a LAN or WAN, or as a fax image.

The system has proved particularly valuable for library use in the transfer of material from rare books and documents. An A4 page generates about 4 million bits of data, although the resolution can be increased so that 10 million bits can be produced.

BT Bureaufax

Finally, this is yet another BT service mainly aimed at overseas connection for companies that do not have fax, or who want to send fax to a company that does not. The main users, however, are companies whose fax is not compatible with that of the intended receiver.

Documents can be posted or hand-delivered to over 50 'Intelpost' centres, or faxed to a central transmission centre and the various rates and charges depend on the service wanted and the country to be sent to.

A new incentive is the offer to reverse the call charge if you are faxing 10+ pages for transmission abroad, a considerable cost-saving if you are not within local charge reach of a centre.

Further information: Bureaufax, Cardinal House, Farringdon Road, London EC1M 3ND

Fax directory

British Telecom publish a fax directory which is available from: Freepost BT1 FAX, BS3333, Bristol, BS1 4YR.

Chapter Six

On-line database access and viewdata

What is viewdata?

In dealing with data processing, database access, telex, teletex and so on, except when dealing with graphic packages for the representation of electronic office 'database', we have been almost entirely concerned with data in a *character* form, i.e. as digits, letters and a few special symbols such as for punctuation. Fax is different of course, in that we are looking at whole images transmitted as ones and zeros so that the picture as a whole is built up by the receiving equipment.

Viewdata, or *videotex* as it is more often called abroad, provides a means of database access which makes use of many of the available effects that can be provided by a modern VDU, alone or in combination, including colour, a range of type sizes, graphic symbols and shapes, all of which are employed in making database and general information enquiry as easy and attractive as possible by presenting it in page form.

At the same time, it must be realised that not all on-line database access is viewdata. There are many companies such as Pergamon (Infoline), Datastream and Extel who provide a very comprehensive series of data banks where the dialogue is on a line-by-line basis rather than in complete pages. We will come back to these when we look at dialogue possibilities.

In addition, although the original idea of viewdata was mainly 'electronic publishing', almost like newspaper/magazine/encyclopoedia via a

terminal, access to information can be thought of as via a complex message box and, in fact, interactive viewdata includes home shopping and banking. In most applications however, the user is concerned with being able to select from a vast range of different private or public data banks and pick out some information.

First of all, let us state clearly the difference between viewdata and *teletext*. In the UK, Ceefax from the BBC and Oracle from Independant TV are services for special subscribers that give news, general information, program news and simultaneous program subtitles for the hard of hearing. Information is made available as a series of *pages*, each of which shows page number, date and time. The services are provided for the totally non-interactive user in the sense that he does not communicate with a central computer in order to obtain information. Instead, all the pages are sent, one-by-one in a cycle. By entry on a special keypad, the user can select a page and after waiting some time for that page to appear in the cycle, it appears as a static picture on the TV screen where it is regularly refreshed until another selection is made.

Viewdata is often confused with this for two reasons. The first is that the general format of teletext and *Prestel*, the only public viewdata system in the UK (hence the one that most people are familiar with), are similar in the banner heading at the top and the use of colour and characters. Secondly, although Prestel can be accessed from a micro, many people see it by access through a keypad with TV output as for teletext.

The actual access to viewdata is more sophisticated than teletext. Instead of just entering a page number for information, the user goes through a hierarchy of menus until the relevant subject is reached. Also, not all the pages are sent in a cycle to the user. Teletext has a relatively limited number of pages and provided the user will put up with the delay waiting for a particular page to arrive in the cycle, it is not too bad. But viewdata generally is handling vastly more data and the delay would be ridiculous if the whole database was sent as a cycle of pages.

The names used for the services are rather confusing because some people use the term 'viewdata' to mean the same as 'videotex', while others would say that videotex means viewdata *plus* teletext, i.e. services that involve text on a VDU. (For the rest of this chapter, the term 'viewdata' will be used.)

Types of dialogue

In order to provide information in the desired form, it is necessary to hold and handle a certain amount of formatted data and we refer to various kinds of *frame* such as those for information, for data collection and those for basic messages (mailbox). Sometimes a frame may be held on disk as a complete page, while others act as an 'overlay', i.e. literally a presentation framework for a page. Yet others need to be dynamic in that they need to allow for continuously changing data such as the current time.

This is further complicated by the fact that graphics features may appear as well.

Data collection frames are basically forms on the screen and must be designed with the user in mind. The entry of data is a kind of dialogue that minimises the chance of errors so frames will have protected areas for the system to supply data and data entry areas for the user. As data is entered, the system may carry out appropriate validation, i.e. produce suitable error messages when data is wrongly entered.

A typical screen for ordering goods might look like

that shown in Figure 6.1. This user screen has three aspects: the top contains general information, the bottom shows how many lines have been ordered and the running value of the order entered, and the middle window allows the actual order to be placed. Items in square brackets are where the user can make entries and the curly brackets show where the system is giving you further information, such as the description corresponding to a particular part number. Also, error messages are shown such as entering a non-existant product code or an out-of-stock item. After entering all the desired details, function key 2 initiates the prompt for credit card number and PIN, after which the total order value is shown and the order is processed to your account.

Don't forget that dialogue also involves output, Figure 6.2 shows a typical screen display, a daily sales report from Effem.

A hierarchical menu is probably one of the most common bases for viewdata dialogue whereby increasingly more detailed access is obtained as you go down various levels.

The following dialogue might apply in a system used by a bookseller. The first three lines classify the material required and then form-filling provides the fine detail.

First selection

1 = Science 2 = Arts 3 = General ...
Subject [ ]

Second selection

1 = Book 2 = Magazine 3 = Audio tape
4 = Video tape ...
Medium [ ]

Third selection

1 = Private 2 = Publisher 3 = Government ...
Source [ ]

AUTHOR [   ] TITLE [   ]

PUBLICATION DATE IF KNOWN [/ /]

Most public and private viewdata systems use a menu approach (see Figure 6.3).

```
     DATE                        SUPPLIER NAME                        TIME

   ENTER F1 TO CANCEL CURRENT LINE
   ENTER F2 FOR PRODUCT DISCOUNT DETAILS
   ENTER F2 TO SEND OFF ORDER AND END RUN


    ENTER PRODUCT CODE  [XXXXXXX] {     DESCRIPTION    }

    ENTER PRODUCT PRICE [XXXXX]    {   ERROR MESSAGES  }

   ENTER ORDER QUANTITY [XXXXX]    {   ERROR MESSAGES  }

                PRODUCT ORDER VALUE £XXXX.XX

   ENTER REQUIRED DELIVERY DATE [XX/XX/XX]

   ENTER CREDIT CARD NUMBER [XXXXXXXXXXXX]

   ENTER PERSONAL IDENTITY  [XXXXX]



   NUMBER OF LINES ENTERED XXXXX
   TOTAL ORDER VALUE £XXXX.XX
```

Figure 6.1   *A conceptual ordering screen*

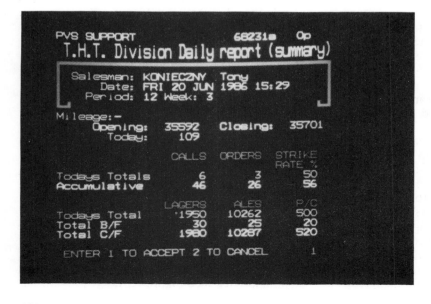

Figure 6.2   *An Effem screen, daily report summary*

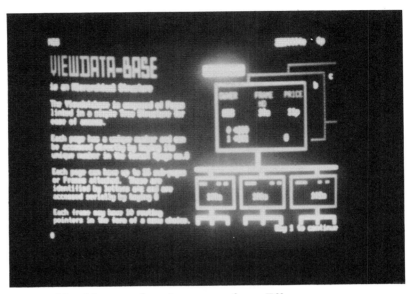

Figure 6.3 *Menu explanation screen from Effem*

Sometimes it is convenient to be able to repeat an earlier page shown or even to browse backwards. Another requirement is to be able to 'escape' from a lower level back to the higher-level menu that got you there.

When the user is accessing data, menus are very convenient and easy to use, but many companies offer specialised data banks which are geared to textual information and there are so many variables that a menu approach would be totally useless. Another aspect is that the user often does not have complete information and what is needed is a database enquiry-oriented language as discussed in the first chapter.

An example might be looking up a court case. You know the year of the action and you think it was held in the Surrey County Court in Guildford. You know it involved Smith Ltd and you want further details. After some menu selection to give you the general service, perhaps: 1 = pre-1960, 2 = 1960–1969, 3 = 1970–1979, 4 = 1980– and 1 = Civil, 2 = Criminal, 3 = Maritime, it would be nice if you could now enter just: Guildford County, Smith *v* ? The system would interpret this as a request to search for and display all cases in the decade specified, held at Guildford with Smith as the plaintiff. From this list you could pick out the one you wanted.

Similarly in a library information system, when looking for subject references, the dialogue must allow for incomplete knowledge and the following kind of inquiry is possible:

Magazine article, 1981–83, UK, "viewdata", "pharmaceuticals","?Manchester?",..

This is asking for details of any magazine article, published in the period specified, dealing with the application of viewdata in the pharmaceuticals industry. The only other help you can give the system is that the word 'Manchester' appears in the title of the one you are looking for.

Computer types call this kind of dialogue *keyword-oriented*. When the enquiry is entered, the receiving computer treats it in a way similar to that when a programming language is converted into machine code for execution. It is looking for certain words in a certain structure. When it has

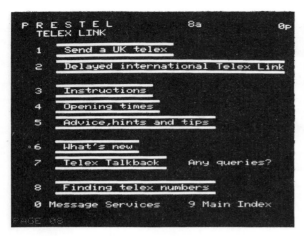

Figure 6.4 *Prestel telex facilities*

of short-circuiting of the menu structure, which can lead to very much quicker information retrieval.

Viewdata features

The main features of viewdata applications whether Prestel-based or private, are as follows:

● Wide range of data, either of completely general use, e.g. Micronet from within Prestel which gives information of value to thousands of microcomputer hobbyists or Prestel itself or Dialog which provide information on just about anything to just about anybody), or in particular for a limited but large number of users, such as the corporate viewdata systems installed for Thomas Cook, ICL, Leyland, etc.

● Volatile data, i.e. data that is constantly being revised and updated, such as share and commodity prices.

● Fairly standardised, easy-to-use commands so that users can flip from one service to another without too much need for further training.

● Fairly standardised screen output formats so that users quickly gain familiarity.

● Many users with differing requirements, possibly widely spread.

● Fast response (or as fast as possible depending on the complexity of access to the data).

● Variety of terminals for access (including home use).

● Variety of access methods.

● Variety of sources of information. Some information providers are often highly specialised in what they offer (as you will see from the examples given shortly), but many are wide-ranging and are accessible from Prestel, which itself is very comprehensive anyway.

worked out what you want, it sorts out a series of accesses to the database in order to extract the required data. Obviously, the more information you provide, the easier will it be for the system to find it. Since you are paying for telephone connect time, it will help you as well because it will reduce the time you are linked, waiting for the system to assemble your answer.

The dialogue is designed with this in mind and often has 'wildcard' options, where you supply part of the information, such as a word from a title, or a range of values, such as year of origin, size, etc. This kind of facility is not common in viewdata, but some private systems allow this kind

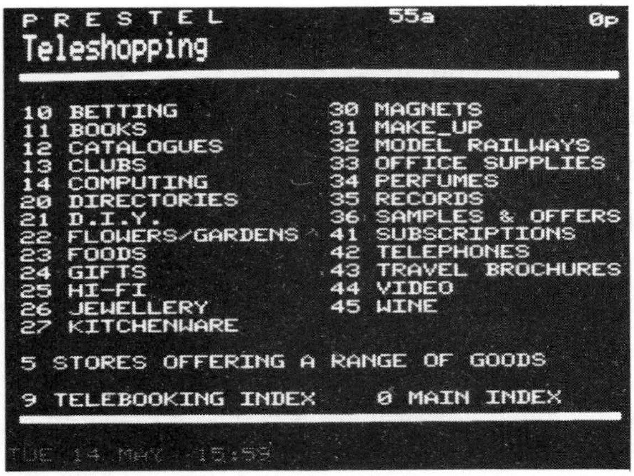

Further considerations that apply to companies planning to lay on corporate viewdata are the possibility of using existing network hardware and other facilities and staff not having to rely on the

Figure 6.5 *Prestel teleshopping*

telephone (or telex/fax/teletex) for up-to-date information. In fact, the larger viewdata users will usually provide a Prestel link which opens up a gateway to many public, specialised data banks. So within the organisation, as well as having the company data bank on-line, marketing people can get at industry sales figures, engineers and scientists/technologists can access data relevant to them, etc.

Prestel

Prestel, whose standards and protocols (with some modifications), have been adopted almost as a standard in their own right, was developed by the Post Office as a means of providing for the private user, a large set of interactive but 'static' screens, in the sense that unlike teletext, the user has far more control on what he can display, but is only really interested in getting information from the 'information provider', i.e. the companies who are in the business of selling information via viewdata services.

Business needs are more demanding in that the interactive nature must be extended so that 'data processing' is possible as well. If a user is already making use of a terminal for order entry and other transaction processing, for viewdata to be really useful, it needs to be integrated with the more conventional data processing activities.

Once corporate users such as the City of London institutions and the holiday companies appreciated that viewdata could provide an effective means of transferring data and information around the company, it started to take off, first as Prestel closed user groups (CUGs), then as private viewdata networks.

Prestel has a number of central computers (Update centres) where the data banks are stored and kept up-to-date. A number of satellite units (Retrieval centres) pick up user commands and relay the relevant pages to the user.

Both Prestel and private viewdata systems take some hefty computer processing to back up both the communications and data storage/access aspects and companies such as Digital Equipment Corp, IBM and Honeywell feature heavily. (Other manufacturers of small mainframes and minis are involved including Data General, Hewlett Packard, Texas Instruments and Univac.)

Prestel itself has about 60,000 subscribers, which is rather lower than was originally hoped (current estimate – 100,000 viewdata terminals). Possible reasons for this are:

● The cost of the service.

● The cost of the terminal adaptor.

● The fact that for use at home, the TV set cannot be used while it is tied up with viewdata display.

● Poor quality – shapes and characters are often made up from blocks.

We must reflect on whether the user actually wants all the fancy graphics of viewdata or perhaps just plain readable text, i.e. non-viewdata information formatting or perhaps just teletex. There are conflicting views on this. However, most corporate database access seems to look like Prestel.

The private user also has the restriction that while using Prestel, both his telephone and TV set are tied up and cannot be used by the rest of the family.

Another factor to be considered is whether a simple numeric keypad, an original Prestel requirement, is sufficient for most people's needs if interactive working is wanted.

The French *Teletel* system is proving much more popular with the public. They have fairly recently introduced a scheme whereby the terminal is given away free with the service, the projection being that the cost will be more than made up by the increased use of Teletel.

The videotex standard laid down in September 1983, by CEPT, the European Post and Telegraph Committee, is a development of the original Prestel and Teletel and it specifies alphamosaic display with dynamically redefinable colours and dynamically redefinable characters, etc. A wide range of characters is required to accomodate Spanish, Scandinavian, German, etc.

Figure 6.6 *Two different keyboards, two different ELF displays (courtesy Easydata Ltd)*

Prestel user dialogue

The general format of Prestel screens is to show the name of the information provider as part of the 'banner' at the top left and the number of the page currently being viewed on the right (a letter after the page number shows the 'frame' number'). Instructions to Prestel are simplified so that a modern telephone-like keypad is all that is needed, including 0 to 9, * and #, where *n* (*n* is a number) selects the *n*th choice from a menu, and *n*# (where *n* is a page number) selects the page bearing that number.

The # character links to the next continuation frame for the page since a page may be several screensful (up to 26, from a to z). (A frame may have associated with it a further menu with up to 10 choices on it.) For example, the Bank of Scotland screen shown in Figure 6.7 is frame *e* of page 3951131, presumably a long statement of a busy account.

The symbols *# give a re-display of the previous frame shown. (Prestel allows this to be repeated up to three times). If there have been any transmission errors and it has been corrupted, *00 is used to re-dislpay the current frame.

Bearing in mind that a dynamic database, by its very nature, may be changing all the time and you may wish to repeat the current frame, *# pulls this from a buffer within your terminal. If it has changed since then, *09 will display the most recent version, by accessing again from the central computer.

The current entry is cancelled with **.

Viewdata access routes

Depending on the application, the user can have access via a specially-designed TV adapter (Prestel), a personal computer with an

```
Bank of Scotland        3951131e         0p
  Statement    Account No 00428407
  Date    Details      Amount      Balance
07Jan85 Div. I.C.I.    22.17        -21.75
09Jan85 Giro credit    18.00         -3.75
11Jan85 Salary        550.23        546.48
14Jan85 456123        -11.99        534.49
15Jan85 456124         -5.99        528.50
18Jan85 456125       -452.23         76.27
21Jan85 Keycard       -55.00         21.27
    This is an example of the statements
    which are presently available to our
    Home & Office Banking users.
  Key 0 Bank's Index,    Key 7 to continue
PAGE 05
```

Figure 6.7 *A statement from Bank of Scotland 'Homebanking'*

radio is starting to be used, especially in areas where a cable is impractical – such as on board a sea-going vessel. There is also some sign of a trend towards the use of private cable such as is used for cable-TV.

VDU display techniques

The simplest form of display is based on the common characters and shapes formed from these, particularly since many VDU terminals have an option to display characters of varying size. Note that as well as sending a character or shape from the central computer to the terminal, it is often necessary to send extra information about the character – which colours, is it flashing or underlined, etc? Where these attributes are sent along with the character, we refer to *parallel* transmission. The alternative, *serial* transmission is to send the character, then the attributes as another 'character', so that they take up space in the transmitted stream even though they do not actually appear as viewable characters. This difference can be shown conceptually as in Figure 6.9.

appropriate (1200/75) modem or a special-purpose terminal, often with autodial facilities (see Figure 6.8).

The actual transmission path can be through the PSTN, leased line or the PSS (Prestel, Prestel gateway or private packet network), but cellular

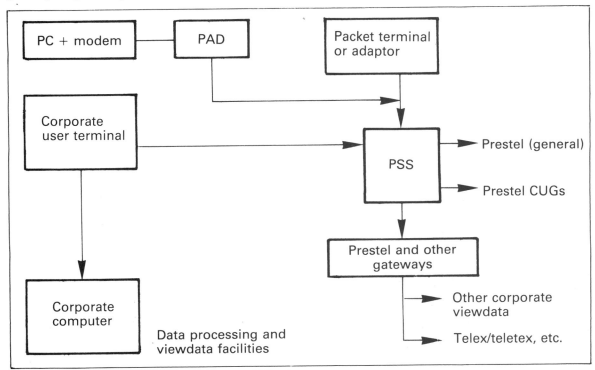

Figure 6.8 *Access to viewdata (courtesy Effem)*

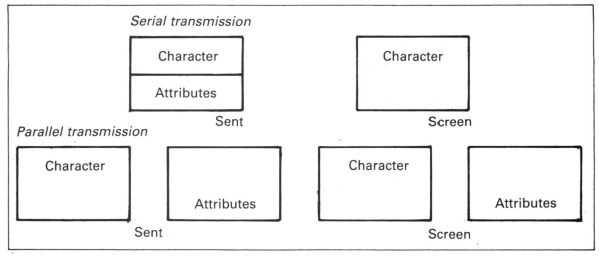

Figure 6.9 *Serial versus parallel transmission of attributes*

Alphamosaic as with Prestel is the next way to build up characters and shapes. Each character has a set number of cells on the screen from which a character is made up. A typical character matrix would involve 7×5 cells (see Figure 6.11)

For graphics, a smaller matrix is used to build up characters. This is based on six cells, two across and three down and using groups of these, shapes can be made up.

Alphageometric is much less restrictive in that shapes are made up from a range of smaller shapes which are sent as *Picture Description Instructions (PDIs)* defining the size and shape of the items to be displayed. For example, to produce an ellipse,

the parameters sent would be the centre, the major and minor axes and the fill colour. The higher the resolution of the terminal, the better is the picture. These days, even small microprocessors can acheive these effects – if your son or daughter has an Amstrad, Commodore or similar machine, the Basic provided will almost certainly have commmands that allow shapes to be produced as part of the language. If you want to print text, you say something like:

PRINT "GOOD MORNING DAD"

The Basic language with the Sinclair QL has many

Figure 6.10 *Shopping with Harrods via Prestel*

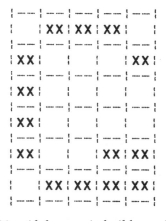

Figure 6.11 *Alphamosaic build-up of the character 'G'*

Figure 6.12 *Using graphics for effect 1:*
a Micronet adventure

Figure 6.13 *Using graphics for effect 2:*
Mars European centres

statements for producing shapes, for example:

PRINT "THIS IS A CIRCLE"
CIRCLE 100,100,50

would display a circle radius 50 near the top of the screen.

Alphageometrics are used in the Canadian public system, but do not have wide application in Europe.

Alphaphotographic With this, characters are made up from a large matrix of dots which can lead to very clever shapes and effects. It also leads to a time delay. The more sophisticated the actual display, the more information that needs to be sent and the more work the terminal processor needs

to do. This kind of output will presumably become more viable when digital networks are more firmly established. BT have been carrying out quite a lot of research and there is talk of 'picture Prestel' which, it is claimed, will provide very high resolution display, once the user has access to the high-speed transmission facilities. These are needed since this high resolution requires about 2 million characters per frame.

Mullard produce an integrated circuit for use in viewdata terminals. It can handle all the effects in the CEPT recommendation and the ROM supports all of the following effects which will give you a good idea of the kind of 'tool-kit' needed by a viewdata dialogue designer:

● *Colour* Up to 31 foreground and background colours ('ink' and 'paper') can be used at any one time, selected from a range of 4096.

● *Character size* Double height, width and size can be mixed up with 'normal' size at will.

● *Character underline*.

● *Reverse* Foreround and background colours can be flipped.

● *Separate 'windows'* This can give several nice effects. One is to have the screen divided into two or more separate areas which can be used for different purposes. Alternatively, a 'box' could be opened up in the middle of a scene to show text relating to the scene. Another possibility is for independent 'scrolling' where the text in one window can be rolled up and down independent of what else is happening on screen.

● *Flashing* This can be a character or set of characters which flash because the pen and ink colours are continually flipped. Alternatively it could involve a third colour. Another possibility is to use three stages of flash so that movement can be simulated. The technology behind such a chip is way beyond the scope of this book, but with these facilities available in a terminal, it is little wonder that viewdata screens can look so effective.

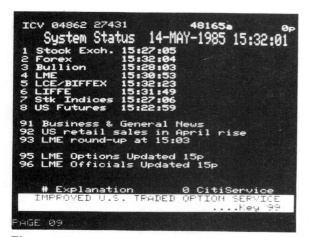

Figure 6.14 *An ICV financial menu*

Viewdata services

The following are a small selection of the services available through Prestel, either for general access or as closed user groups.

Dialog The world's largest database access facility giving access to a range of 300 different main databases on an almost encyclopoedic range of subjects. It is based in the USA. A very small selection of Dialog databases includes the following:

American Men and Women of Science
Child Abuse and Neglect
Food Science and Technology Abstracts
Harvard Business Review
Trademarkscan
World Affairs Report

Citiservice Jointly operated by Prestel and ICV Information Systems and providing stock exchange and financial information.

Fintel From the *Financial Times*, providing foreign exchange and financial information.

Nexis A full-text news database covering US papers such as the *Washington Post*, *New York Times*, *Congressional Quarterly* and a range of UK services. A similar database called Lexis provides legal information. (Until recently, there was a comprehensive legal service provided by Eurolex which is no longer available).

World Reporter Another news database that gives viewdata access to *The Guardian*, *The Financial Times*, *The Economist*, *Washington Post*, BBC World Broadcasts and External News abstracts and various wire services such as Associated Press.

Scicon This provides access to POLIS, the Parliamentary Online Information System detailing all government and official parliamentary publications. It also gives access to

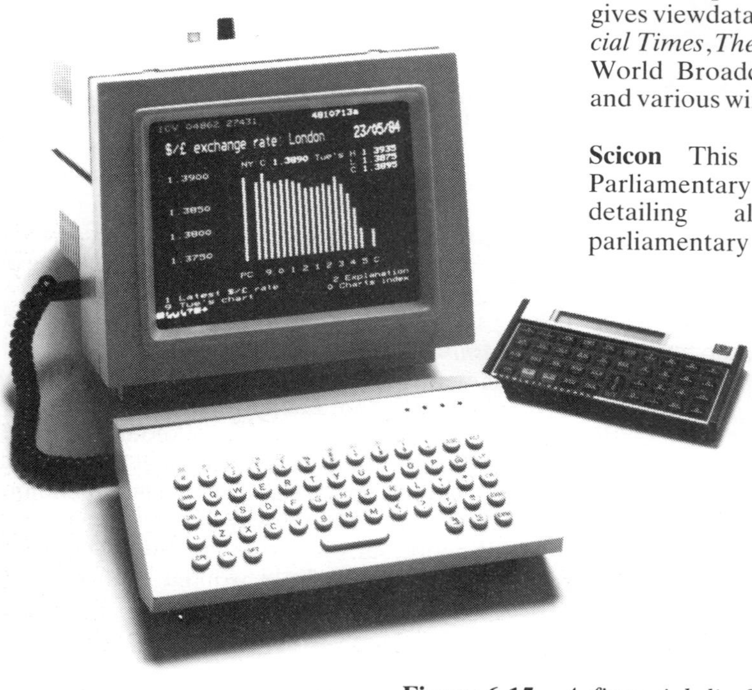

Figure 6.15 *A financial display from ICV on an Easydata ELF*

DHSS library information through 'DHSS-Data' and 'Acompline' which is geared to the GLC research library.

Blaise This is operated by the British Library and provides very comprehensive bibliographic reference to books, magazines and journals.

Datastar This has an assortment of 80 odd databases covering financial, business and medical information (including the British Medical Asociation Press Cuttings File).

Pergamon Infoline A wide range of business, technical and scientific databases, including *Who Owns Whom* and Dun and Bradstreet's *Key British Enterprises*.

Context This is a private service for architects and builders provided by The Royal Institute of British Architects and AVS Intext.

Infotext Another private service, for the film and TV industries.

UAPT Yet another service supplied by the United Association for the Protection of Trade, it provides on-line credit-checking on up to 50 million names extracted from voting lists and county court data on bankrupts and debtors.

Other services are provided for more general use such as:

The Hackney Bulletin This provides a wide range of information on facilities in the London Borough of Hackney.

Berkshire County Council A service for Berkshire employees from Shire Hall in Reading, it is installed in libraries and educational institutions.

Cymrutel This offers a local educational service in North Wales.

Amongst other companies, Curada Associates publish a Directory of on-line Databases which is updated twice a year.

HOBS

This is a service, already referred to, supplied by The Bank of Scotland through Prestel for private and business users. The main menu shows the following:

- Account details and statement.
- Cash management (business accounts only).
- Examine regular payments.
- Change password.
- Service requests.
- Make bill payments.
- Inter-account transfers.

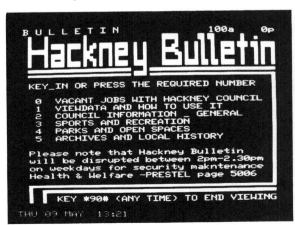

Figure 6.16 *Welcome from the London Borough of Hackney*

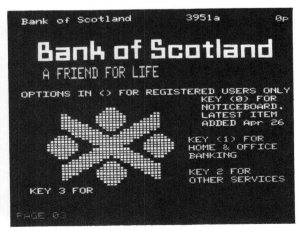

Figure 6.17 *The HOBS 'welcome' screen*

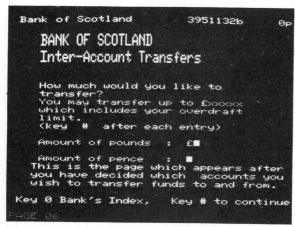

Figure 6.18 *HOBS inter-account transfers*

Using the on-line facilities, you can examine your account (up to 250 transactions), obtain an analysis of standing orders and direct debits (when due, date of final payment, etc.), transfer money electronically (Prestel provides a list of other subscribers to HOBS and people who will accept this form of payment), transfer money between different accounts (such as from current to investment).

Users of HOBS need to subscribe to Prestel for access to HOBS services. The company is currently offering a special price of under £100 on their recommended (Tandata) HOBS Prestel TV adaptor. For application to the HOBS service, contact Department HOB/CLD, Bank of Scotland, PO Box 403, 2 Bankhead Crossway North, Edinburgh, EH11 0NU.

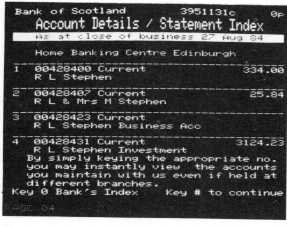

Figure 6.19 *HOBS statement analysis*

Private viewdata or prestel?

The options available to a potential user of viewdata are Prestel itself, closed user groups within Prestel, Prestel as a gateway to other viewdata services or closed-network private viewdata. When making a decision, assuming that the cost factors have already been defined, it is worth looking at just a few of the advantages of a dedicated service.

Generally, any purpose-built system will be more appropriate than one tailored in. Then, the number of frames available will be greater than supplied by the information provider since it will be up to you to decide on this. There can also be better editing facilities and better security measures (to your own requirements) even if they are not so user-friendly. Form-filling dialogue can be used extensively for data entry: this is unlikely to be the case with public viewdata.

Since you are responsible for deciding what you get, it should not be very difficult to include electronic office software within the viewdata service. It also means that you can take advantage of changing technology such as will come with ISDN.

Corporate viewdata systems

British Leyland (as it was then) set up a viewdata system in the late 1970s called 'Stock Locator' to help car dealers track down different models sitting unsold in other showrooms. This has recently been substantially upgraded by the company Istel to a system called 'Dealer File'.

As well as locating stock, the system deals with orders/sales and drastically reduces the paperwork that a dealer is involved with. The system has a number of components: 'Profile' shows a list of enquiry contacts for dealers; 'Finance and stock control' locates vehicles and transfers payments electronically; 'Service' deals with warranty claims and guarantees; 'Sales/Order processing' provides national sales statistics and financial information relating to sales and targets. The system is backed up by Istel's data communications network which has over 60 access points throughout the country. Ten large mainframe IBM and Amdahl computers provide the main

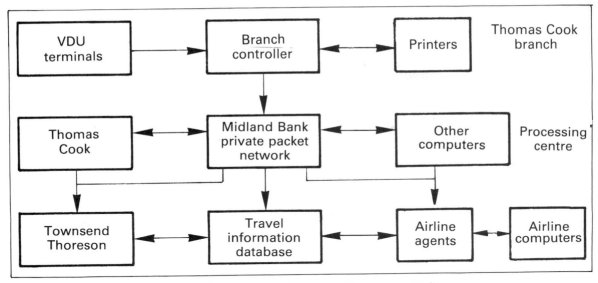

Figure 6.20 *Thomas Cook viewdata structure (courtesy Microscope Ltd)*

processing facility for Leyland and data is switched between them and the Ten Digital Equipment Corp VAX computers which support Dealer File. General Motors have a similar system which handles ordering and location of 100,000 different stock lines, in all over 37 million items.

Microscope, a company based in Maidenhead, have installed a number of viewdata systems both in the UK and abroad including systems for ICL, British Rail (Nottingham and Marylebone) and for the travel agents Thomson Holidays and Thomas Cook. They have also put in national viewdata systems for the postal authorities of New Zealand, Norway, Denmark and Belgium. They claim that in the period July 1982 to April 1985 they have installed £2.5 million of systems with over 6,500 ports in total. The Thomson system handles a large number of terminals which make about 155,000 calls per week. In its first week of operation, it dealt with 350,000 calls. About 70% of their holiday business is handled on an on-line basis. The system is based on Microscope's *Videogate* concentrator.

The Thomas Cook system is even more complex (see Figure 6.20). It integrates viewdata and computer-to-computer communications and the facility is provided from all the branches using Videogates. Each branch has a control unit linking the various VDUs and printers to the company

mainframe. In addition, databases provided by the airlines and Townsend Thoresen can be acessed to organise and book travel schedules. The whole system is geared to the private packet-switched network provided by Midland Bank.

Another recent installation has been for Berni Host (Grand Met group) to handle accounting, stock ordering and delivery, and labour data for pubs and bars in the group, which includes Chef and Brewer, Berni Inns, City Limits and others. It is estimated that about 3000 people within the group make use of the facilities.

Autophone, whose Talkback system was mentioned in Chapter 4, are also involved with viewdata and they have installed a system for Audi Volkswagen. It is called *Dialog* and links 380 dealers with the Autophone centre in Milton Keynes. The user company felt that the user-friendly viewdata style of communication with the computer would be liked by dealers and would allow for upwards expansion.

The workload specification was for about 600,000 transactions per month and the very high speed data transmission requirement precluded the use of an existing viewdata system. Autophone in collaboration with Bishopgate Computing, had already installed the *Topic* viewdata system for the Stock Exchange and several features of this

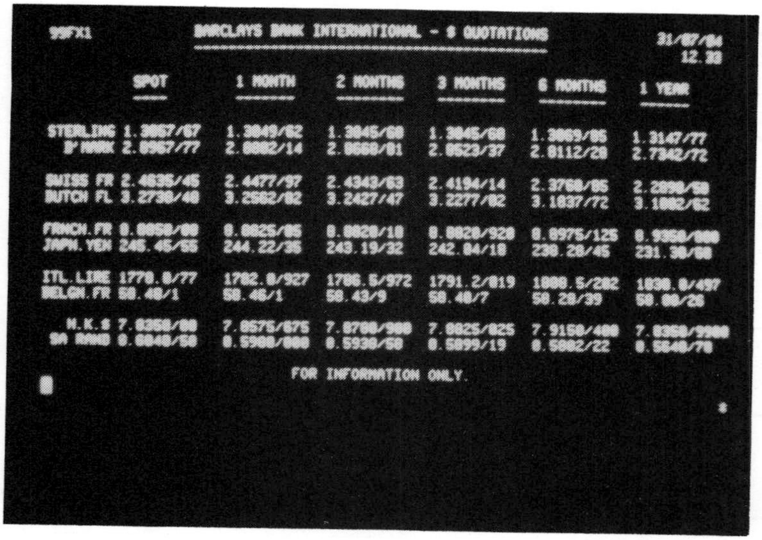

Figure 6.21 *A text display of dollar quotations from Datastream*

recommended themselves to Audi, in particular the facility for off-line data collection and bulking for later high-speed transmission.

The system was implemented over four years ago and dealers can interrogate and update the database getting immediate response to enquiries such as the location of particular models of particular colours. This is obviously a very impressive demonstration of efficiency to the potential customer. Standard or special orders can be placed and delivery dates pulled off. Another possibility is the exchange of vehicles between dealers. Dealers claim that the system improves parts ordering. Conventional mailing costs disappear and by taking advantage of the bulking facility, cheap-rate phone time is used.

Datastream

Datastream is a City-based company, owned by the US Dun and Bradstreet Corporation. It provides non-viewdata, financial on-line services in a wide range of areas to support investment and other decisions by brokers, dealers, bankers, accountants, economists and others.

Information stored covers UK and international markets and includes: shares, commodities and stock market indexes; property and bonds; money

markets – exchange, interest and deposit rates; corporate profiles and financial statements; economic and industrial surveys such as IMF and OECD.

Databases are updated by on-line connection with international investment markets, collection of magnetic tapes from leading agents and vendors and manual collection of prices, rates and statistics from published sources.

For up-to-date sterling and dollar quotations, Barclays Bank and National Westminster Bank on-line services are accessed. When the customer wishes to display data, he can make use of a range of software tools including a program called *Minder* which can display up to 30 different values formatted to your choice and showing the date/time of the last update.

Investment research based on this kind of information can be used for share transaction planning, dealing in futures and examining companies for mergers and take-over analysis. For fund managers, automated fund management is possible including accounting for purchases and sales and a means for performance measurement. The other aspect of Datastream is that it allows you to build up your own databases on their computers and carry out projections and forecasts on your data.

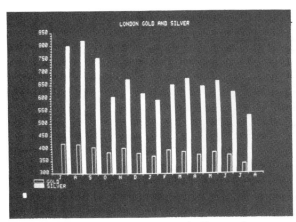

Figure 6.22 *London gold and silver prices from Datastream*

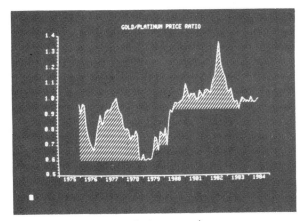

Figure 6.23 *Gold/platinum ratio*

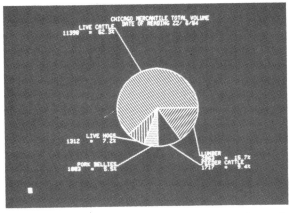

Figure 6.24 *Data from the US 'meat exchange'*

Access to Datastream

For the large user, the company can provide computer-to-computer linkage based on IBM protocols over dedicated lines. The small user can make use of a PC (with colour if required) or take advantage of the cheap offer from Datastream of a Lynwood Alpha Graphic intelligent terminal, or a Lynwood GD1 dumb terminal. A 180-cps printer can also be supplied.

Single-terminal access can be PSTN (modem or accoustic coupler) or PSS and with an intelligent terminal it is possible to 'download' Datastream data for use with your own software, such as Lotus 1–2–3.

The large user can make use of 'batch update' facilities whereby, instead of entering data through a keyboard, you can arrange for your computer to send it from a high-speed device, taking advantage of off-peak transmission charges.

The graphics facilities available mean that users can build up files, process them and display results as graphs, bar charts, etc. This can fit in nicely when displaying Datastream-held data or your own and is particularly useful in combination with statistical analysis software which amongst other facilities, allows you to carry out time series analysis and forecasting.

The company has very comprehensive user manuals and provides training courses in operating procedures and specialist application areas.

Effem Management Services

This company is owned by the Mars Group and operates in conjunction with Mars Group Management Services, offering a range of computing and data processing services. In particular, it supplies a viewdata service called *PVSNET* which also allows for a range of data processing activities including data analysis, data capture, frame editing and validation. Using the system, a company can set up its own viewdata service including electronic mail, based on the Texas Instruments Business Systems range of computers used by Effem or its own similar hardware. Electronic mail allows users to receive a summary

143

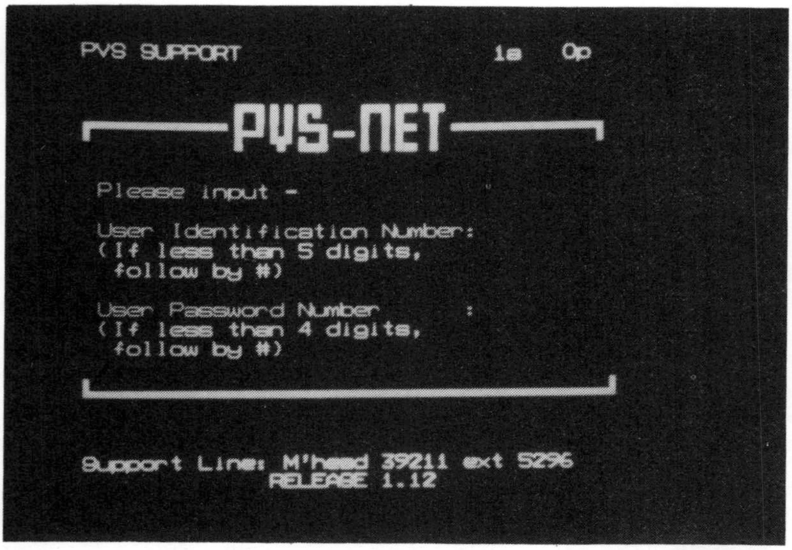

Figure 6.25
Log-on for PVS-NET

of what is in their new mail tray and on file. Other options can be tailored for a customer. An extended electronic mail facility allows for multipage messages, directory searching, open broadcast, and message priority and redirection.

Another major facility is *TRANSVIEW* which allows for file transfer between intelligent terminal and host using special error-minimising protocols. The main application is for large organisations who support a number of software development centres, each developing applications programs. Using PVSNET, a 'central register' of programs can be maintained and displayed. If a program is wanted, it can be selected and transferred to the terminal just like any other file. This means that users and software developers are aware of what is happening, so effort is not duplicated and perhaps more important, all users work to the most up-to-date program versions.

Using TRANSVIEW running under PVSNET, users can make use of the intelligence and storage capacity of their terminal to set up and validate data off-line from the main system. Then, when convenient, the data can be sent to the central computer for overnight processing. At the same time, up-to-date prices, customer credit ratings, etc., can be sent to the terminals so that the 'shop floor' is working to the latest figures.

This kind of facility could be obtained via a gateway from Prestel, but the actual mechanism is quite clumsy. From Prestel, you access a page which provides access via the gateway. This picks up a link to the host computer you want. Data entered by you is in frame form and is sent as a record to the host which is running the appropriate applications program. The output from processing is returned as a record, to Prestel, which then presents you with your display frame.

Although you may believe that you are carrying out batch processing in that, for a particular session, you may enter several transactions, this approach imposes an on-line approach because you have to pay for all the connection facilities – every time Prestel sends your input to the host, the host operating has to recognise a processing request, call up the relevant program, run it and then send back the results.

With PVSNET you are, in effect, directly on-line to the host computer with the advantages of lower response times and viewdata graphics formatting.

Yet another advantage is the ability to mail reports, analyses, etc., from terminal to terminal by file transfer. Terminal access is available via microcomputers, TV adaptors or Texas Instruments, high-resolution TIVIEW-3 terminals

Figure 6.26
Effem electronic mail

which are CEPT standard. The terminal can be fitted with a Tandata modem with autodial facilities which allow numbers to be dialled via the keyboard, or an external 100/75 dial-up modem can be employed; 1200/1200 is also supported.

Information frames on the terminal are 22 lines deep with either 40 or 80 columns depending on the print size and can be selected by Prestel-like dialogue and by menu.

It has a number of facilities that are very convenient for the viewdata user such as providing the ability to be able to dump frames to a printer and save frames to disk for later, off-line display. A 'carousel' of stored frames is also possible in which a set of frames can be displayed with a specified delay between them. This is particularly effective for PR and news in reception areas, exhibitions or even shop windows.

Once you have set up your database, you have page number access to these and Effem data and you can take off your displays as you want.

A complex system has been developed for the

Figure 6.27 *Philips HCS 110 executive viewdata terminal*

Netherlands Schuitema retail chain. The system is based around PVSNET, TIVIEW terminals and TRANSVIEW, which is the software for the error-protecting protocol used to transmit files via viewdata format. Terminals in the stores are all to some extent intelligent and consist of cash registers (with optional bar-code readers), weighing machines, hand-held terminals and employee time recorders. A large store may have a mini-computer handling several PCs, while the smaller stores may have just one PC with a regional minicomputer to handle several of them. All the PCs and minis link to a central mainframe which provides prices, descriptions, weights, packaging method, etc.

Store terminals are linked to an intelligent mux which has memory associated with it. Bar-code data is bulked on 'external memory'. The mainframe updates the mux with amendment data although the stores PCs will, of course, update the mux as well, as transactions are entered.

The PCs collect programs by file transfer from the mainframe and run them as required, bearing in mind that the users generally do not have computer expertise. This is why the viewdata is so friendly as well as being flexible.

Users can obtain graphical statistical reports and with one key depression, can flip graphs from line to bar chart to pie chart.

The local programs carry out file management (updating master files on the PC with mainframe or mini updates and vice versa), calculation (various programs to work out meat prices, sales margins etc.) and analysis (to analyse sales by various items and groups, theoretical cash takings, productivity analysis from cash registers and employee time recorders and many other analyses that are of value to the store manager).

The Philips HCS 110

To round off this chapter we should mention a very neat, purpose-built viewdata terminal from Philips Business Systems. The machine has an on-board autodial modem, 34K of RAM/ROM, a printer and magnetic tape back-up. It acts as a viewdata terminal and can also be used for electronic mail, telex, etc.

In addition, it can link in with videodisk and VCR and can provide a programmable page carousel.

The terminal has been supplied to the Halifax and Abbey National building societies and to Thomas Cook.

Chapter Seven

Local Area Networks (LANs)

What is a Local Area Network (LAN)?

In order to introduce this subject, we must first review what we mean by 'network' and 'local area'. Perhaps the easiest way to think of a network is as an arrangement of input/output devices, transmission paths and a control mechanism that allows for the *inter*communication and *intra*communication of information and/or data.

Thus, we have the public switched telephone networks as provided by BT and Mercury, which allow for voice or data transfer between subscribers and which have gateways to the international networks, telex, etc. Chapter 3 mentioned the private telephone network used in the London Underground system based on Thorn-Ericsson switches, which has a link to the PSTN. The packet-switched system provided by BT (Packet-Switchstream or PSS) can be thought of as a network within the PSTN or a network that has links with the PSTN.

We have looked at various transmission media in Chapters 2 and 4 and it should be apparent that radio and TV are also networks, although mainly for one-way communication.

However, all the networks we have considered are intended for *wide area* use, which means in effect that the distances involved between the various senders and receivers, the switches and the intermediate hardware are largely irrelevant (except for the cost of using the service). For voice, the PSTN can link you to Australia or your next-door neighbour along existing transmission media. To send data, you will obviously need a modem somewhere (unless using a digital network) to convert your digital data into a form that will go down the line.

Whether a network is 'wide area' or 'local' is not directly a function of the distances involved but rather how communication is acheived. This is further complicated by the fact that 'inter-networking' is very common. The can either mean the interconnection of similar or different LANs by means of a 'bridge' or the opening up of a LAN by means of gateways to outside networks such as fax, telex, viewdata, etc. The gateway is essentially an interface between two different services (networks in this case) that resolves their basic differences, such as the protocols under which they operate. (Anticipating the appendix that deals with the ISO 7-layer model, open systems approach, a bridge is involved with layer 2, the data link layer, while a gateway is involved at level 3, the network level.)

Perhaps a compromise 'definition' of a LAN might be something like : *A grouping of digital data communication and communication-management devices that does not make use of an external telephone system, where users are mainly concerned with short bursts of processing and within which, distances involved are probably less than hundreds of metres.*

There are even exceptions to this as you will see later – Ethernet implementations can span nearly 3km and with broadband transmission, modems may be needed. Also, audio and video are playing an increasing part in LANs.

Perhaps the greatest difference between LAN and WAN is the fact that processing is in bursts – there will be very few times when a terminal needs to be linked to a processor or another terminal for more than a short time.

Using this working 'definition', we next need to see why a local-rather than wide-area network is employed; then we can look at the components of a LAN and the various arrangements that are possible.

Why local area?

We must accept now that the vast majority of data processing applications involve some kind of distributed processing network, in which users in many different applications have access to computing power and disk storage in different forms, at different locations within the network. Except for the smaller company, most of the processing is controlled by, if not actually carried out by, a mainframe or large mini, with various satellite minis, PCs and other bits of intelligent hardware supporting the applications. (Chapter 6 discussed the Netherlands supermarket application where each store has a PC or mini to hold local files and to carry out local processing, while a mainframe and various minis ensure that local store data is up-to-date and consolidate all the information so that management can be made continuously aware of the overall company position.)

To set up such a network is not cheap, even if the benefits outweigh the costs. In other words, the user will need to trade off the capital investment in modems, telephone lines, communications software, etc., against the benefits and savings arising from faster processing and more accessible, up-to-date management information.

The small user is less likely to be able to afford the outlay required and will need cheaper alternatives. In a typical organisation it is generally the case that over 75% of the data handled is generated internally, i.e. input, processed, stored and retrieved locally. Less than 25% is entered from outside or intended for external use.

In addition, some applications will not involve great distances. For example several PCs may already be in use within a building or complex of buildings and provided the relevant communications software can be purchased or custom-written, it might be possible to connect the PCs with twisted pairs or coax, using line drivers to overcome the problem of losses along the line. This may be quite satisfactory if all that is needed is the occasional 'chat' between two PCs, where one perhaps sends a Lotus 1–2–3 file or a piece of electronic mail to the other.

But a far more common requirement is for a number of users to want to share files all the time and also have access to one or more printers. In a typical modern office we might have three word-processing clerks generating letters and reports, two managers using spreadsheets to process accounts data and 3 clerks entering order transactions. Let us consider the alternatives for the hardware needed to support these activities.

Suppose each user has his own 'dedicated' PC. The word-processing clerks spend most of their time entering and altering text through the keyboard. All of them will need disk space for the PC operating system, the word-processing program itself, the text files produced and any standard letters, paragraphs, etc., that are available to make the word processing more efficient. Occasionally, they will want to produce a draft printout, to check the contents of a letter for example. Then they will want to generate final, high-quality printout for the customer or their boss.

The managers will certainly need disk space for the programs and the files or databases. They will need access to a printer so that financial and management reports can be produced. They may also make use of a graph-plotter so that they can produce graphical financial displays, perhaps to show shareholders or financial backers.

The data entry clerks are primarily involved with running data entry, validation and update programs and will probably have no need of a printer. But, unless each order clerk is servicing his/her own group of customers and his/her own stock items, life will become very difficult. Presumably, the order clerk will want to do a stock check and a credit check prior to accepting an order. Assuming a common stock file, each order-entry terminal will need an up-to-date stock holding figure and since the PCs are independent, this is not possible, Similarly, what is to stop three buyers from the same company that is close to its credit limit, putting in three orders, any one of which will breach the credit limit, to all three order clerks at the same time.

Unless a crippling work-load is being handled, it is unlikely that the word-processing clerks and the managers will be able to keep a printer running all the time, so it would seem sensible to share printing facilities. Perhaps the company could buy a switch box and connect all users to the draft and letter-quality printers, so that when one wants to print, the box can connect him/her to the relevant printer. OK in theory, but a lot of time will be wasted, especially if the users are in different rooms from themselves and the printers. If a printer is in use, everybody else will have to wait.

Even worse, is the situation with disks. The manager may want to 'spreadsheet' data that has been affected by the order entry, so the files must be transferrable. Similarly, the manager may want some of his financial data to be included in a word-processed report. The idea of passing floppy diskettes around the office is just too horrible to contemplate. Can you imagine the problems? Anyway, what if some of the PCs are fitted with non-removable disk drives? Also, what does an order entry clerk do while the manager is doing things to his diskette on another machine? This kind of system might just work for about three intelligent users, of whom one has the authority to impose a standard approach, but would dissolve in chaos under any other conditions.

For less sensitive situations, where general enquiry and text processing covers most of the work, a number of companies offer complete networks that can cope with quite large numbers of users. For example, Xerox offer their *XC-24* network which can span a distance of 1800 feet with two repeaters. The system supports Xerox 'workstations' or terminals and a basic configuration can have three printers and has special software to make the user interface easy to handle. This includes *Screenmate* which simplifies access to system activities such as file copying. A word processor can also be provided which, within a range of other features, allows for multiple screen windows, automatic paragraph numbering, simple arithmetic and simultaneous editing and printing.

Nestar Systems offer *PLAN 2000/3000/4000* which are systems based on IBM PCs. The smallest can link up to six PCs through the hard disk of a PC XT host. Each PC is expanded by fitting a network interface card into the expansion port and they are connected by coax. The 3000 system is built around a 50-Mbyte capacity file server with streamer tape back-up and supports IBM PCs and Apple micros. The 4000 is even larger and has enormous file capacity (up to 1000Mb). For further expansion, it can be linked into a mainframe system.

London Transport Signal and Electrical Engineers Department have installed a 4000 for budgeting and estimating. This was in preference to making use of mainframe facilities because they felt that they had much better control over problems such as ensuring access when needed, sufficient computing time and minimal down time. They plan to extend the system by means of modem links to other buildings and to access the IBM mainframe used by LT.

Nestar have also installed a 4000 system for the City company Postipankki (UK) Ltd. It supports 28 IBM PCs and 16 Apple III micros linked over all five floors of the building to a central 237-Mbyte file server. Ten of the PCs have access to Reuters for foreign exchange and other trading. Software available includes word processing, electronic mail, various spreadsheets with graphics capabilities and access to financial databases. Reuters themselves also have a Nestar system to handle financial dealing.

In summary, we can say that a LAN may be of value where a number of different functions are carried out with the aid of a common database and/or where a number of different devices of different types need to interface and share each others resources. The selection of a LAN involves a trade-off between limited-distance transmission and reduced costs on the one hand and more sophisticated application software and higher transmission rates on the other. However, the network must be structured in such a way that all users get a reasonable share of resources and it must also minimise the effects of hardware failures.

In addition, consideration must be given to disk access times since no matter how cleverly the LAN is structured and no matter how fast the data transmission rate, if many users need to access disk, the overall efficiency will be reduced. Some suppliers have introduced additional RAM, called *cache memory*, which in effect acts as 'silicon disk' in that often-used data can be stored in it, thereby reducing the need to access disk. Access times to cache memory can be thousands of times faster than to disk.

Yet another consideration applies to disk files. There needs to be a protection mechanism that prevents two users updating the same disk area. There is no problem with reading – two word-processing clerks could both be copying the same standard paragraph from a common disk area into the contracts they are preparing, but what happens if two order clerks are trying to update a stock level or customer credit level at the same time while another is making an enquiry on the same record.

The answer is some form of *record locking* which means that the software will prevent read (enquiry) or write (update) access to a record or disk area while it is being updated. The potential user of a LAN must look at this very carefully, since some suppliers will claim that record locking is taken care of whereas the whole system may lock as soon as one user is accessing the disk. This is a possible answer if the usage of the LAN is fairly low, but in general is a very unhappy way to manage file security. One approach is to enable the system to set up disk areas called *volumes* which can be password-protected for 'private access'. Access to the 'public' can be allowed on a read-only basis and if anyone is allowed to write, the system must have some form of *semaphore* capability. This will signal that a particular record or disk area is in use and will cause other potential users to hold off until the transaction is completed. This, of course, means that the system must queue up the requests and have some priority scheme so that a fair distribution of facilities is allowed, while at the same time providing an over-ride for urgent service. The increasing use of LANs is due to several main factors:

● Microprocessor circuitry has advanced so that very sophisticated applications can now sit in ROM and a whole complex of hardware can be attached by inserting a card (circuit-board) in a PC.

● Hardware manufacturers and software suppliers make increasing use of standards and data transfer, and error-handling protocols tend to be built into the network hardware itself, making it possible for the devices involved to concentrate on the application

without having to worry about the network interface and control.

● Coax cable and optical fibre technology have come a long way recently and very high transmission rates are now possible.

● Users have an effective means of getting at computing facilities, under their own control (if properly managed), without having to rely on a mainframe or mini outside their control with shared access to a range of peripheral devices, any one of which might be too expensive for dedicated use.

Components of an LAN

What we really need is a structure that supports common processing power and disk space and the shared use of printers/plotters, together with a control mechanism that ensures a fair access to facilities. A typical LAN will contain a number of user terminals (VDU/keyboards) which might be PCs in their own right, a network controller, a *file server* and *printer server*, all linked via the transmission medium. The network processor will often have some RAM buffer available to it as a high-speed access work area in which to queue messages and data to improve the efficiency of the management and control process.

Not every LAN will contain all of the components shown in Figure 7.1: in fact only the very largest and most complex will. The number of terminals supported will vary from LAN to LAN.

The network controller will decide who gets what and when, within the confines of the design of the network, and handle problems such as device failure. The network processor carries out all the applications program processing needed that is not in the hands of the terminals.

User terminals may have their own disks but, if they want to share disk facilities, the file server must be able to allocate network disk space, ensure that two users are not both updating a disk area at the same time and ensure that network 'system' files are adequately backed up for security reasons. The network management software must be pretty smart to be able to cope with back up of user files, i.e. as directed by the user, as well as

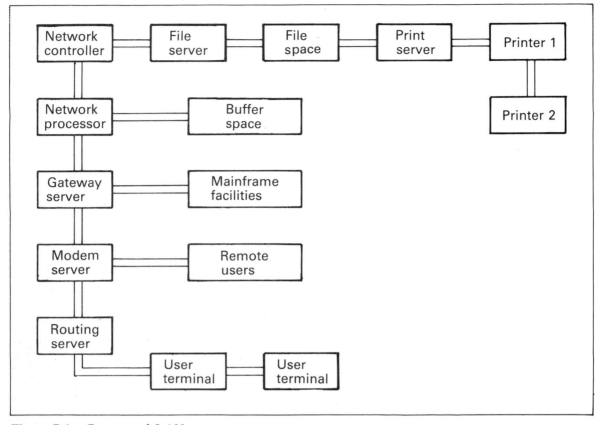

Figure 7.1 *Conceptual LAN components*

be able to take copies of its current management files. This is often a very severe problem and you must satisfy yourself that your dealer can come up with an answer.

The print server must ensure that users can get at the printer in a fair way. When, for example, several users want to print, a queue must be maintained and the relevant print lines 'spooled' to a disk file until the printer is available. The gateway server may be available to link the LAN into a mainframe, using the appropriate protocols. A modem server will enable users outside the LAN to gain access to its facilities.

In some applications a central network processor does all the applications processing – spreadsheet, word processing, database access, etc., in such a way that each user thinks he has his own independent processor. In others, it is more efficient for the terminals to have their own in-telligence (processing power) and the jargon says that terminals are *loose-coupled* if they are intelligent, and *close-coupled* if they are just 'dumb' input/output devices with just enough sense to interact with the network. A majority of dumb terminals raises another contention problem in that the main processor does not have unlimited memory and needs a good memory management scheme to be able to support both network management and control processes as well as the applications programs requested by user terminals.

We will soon discuss the internal working of the possible LAN *topologies* or structures, but with all of them, the network itself is in the hands of a network operating system, usually ROM-based, and however 'dumb' the terminals, they must be able to communicate with the network. A PC makes a good terminal since it has plenty of processing power and disk space for applications and a card holding the communication software can easily be inserted into it.

There is a very tricky problem inherent with PCs and, in fact, with many components of the LAN itself, that the potential buyer must be aware of. The network operating system will often have been designed for a range of terminals and may need tailoring to the customer's PCs by the company that installs it. A specification for the particular PC is therefore needed. Unfortunately, PC manufacturers are always trying to take advantages of technological advances and remove previous deficiencies and bugs, so a new version of a PC may run programs in exactly the same way as far as the user is concerned, but with subtle internal differences that require the software to be modified. In other words, you may have bought an IBM PC together with Lotus 1–2–3 or Symphony and after a while, you decide to buy an upgraded PC. This is likely to mean that your version of the applications software must be 'tweaked' to allow for any internal changes. Fortunately, manufacturers tend to supply fairly complete hardware specifications to software houses and, in their own interests, the software houses try to anticipate, or obtain advance warning of hardware changes. However, if you are putting in an LAN, as with any machine, you must always bear in mind the possibility of future expansion and you must be very careful to ensure that the version of LAN operating system in use and the applications programs are compatible with the new PC .

Media and transmission techniques

The basic transmission media, twisted wire pair, coax cable and optical fibre, were discussed in Chapter 2. All are used in LANs, with optical fibre being used increasingly. Both baseband and broadband techniques are employed. You will recall that with baseband, each node transmits at the same frequency, leading to contention and queuing possibly with the aid of TDM (time-division multiplexing). With broadband, nodes can use different frequencies. so nodes can send/receive simultaneously with the aid of FDM (frequency division multiplexing). A particular channel can be dedicated to a particular source–receiver pair or all channels can be contended for.

Philips Business Communications offer *Sopho-LAN*, which is a broadband tree-structured LAN based on coax. Analogue and digital data can be carried, as can voice and video. The system can be interfaced into their *SophoNET* wide-area network facility. A simplified version for basic electronic office users can be supplied based on a single channel.

In general, twisted wire has a lower bandwidth than coax, which in turn has a lower bandwidth than optical fibre. Twisted pairs are not as uniform as other media, which means that over longer distances, errors and signal losses start to become a problem. Coax is much more uniform along its length and is far better shielded, although the frequency range is not necessarily that much higher than twisted pair. Optical fibre has a very much higher bandwidth than the other media but introduces the problem of how to interface the electrical signals with the light-source that conveys the data into the fibre.

Connectors are easily available that allow for tap-off from coax, but, although developments have been announced recently which enable a light beam to be split with the aid of special mirrors, for practical purposes you cannot split optical fibre.

Note that error-handling procedures and protocols must be built into an LAN just as with any other network. Individual packets that circulate round or through the LAN will contain data for parity checking or CRC (see Chapter 2) to detect corrupted data, but equally important are procedures to ensure that packets have reached their destinations, i.e. packet arrival confirmation, otherwise the sender will not know that he is being listened to.

Network topologies

Almost all works dealing with LANs, bring up this subject sooner or later. However, it is not quite as important as it once was since, although LANs can be structured in a number of different ways, some structures can act like others and the choice is largely decided by the application anyway. As well as needing to talk to the processor, the devices may need to talk to each other and, in general, a choice must be made between the amount of work that the individual elements of the network must carry out in routing messages intended for other devices (distributed network processing) against the complexity of the structure where devices link individually to the processor (localised processing). Each LAN application needs careful design because there are many conflicting factors that affect the choice of topology: transmission medium, control, device access to the network and ordered access to its facilities.

Network control manages the whole operation and may be local or distributed.

Access methods decide which node gets network facilities at any time and allocation techniques decide how much of the network capacity is seized by a node and for how long.

Star structure

In this, the intelligence and control is at the centre and terminals 'radiate' from this centre. The central processor is able to deal with requests for service from any terminal and will usually handle all the communications and applications software and contain all the file space needed. It means, of course, that if one terminal wants to talk to another, it must do so via the central processor. The larger company can take advantage of their PABX to handle LAN switching in a star structure, particularly since the LAN can ride on the existing telephone cabling.

Slough College of Higher Education run an excellent example of a large-star LAN, based not on a microprocessor, but a pair of linked Harris mainframes. The college is divided into a number of departments housed in separate buildings, all of which have various computing requirements. Engineering need to teach elements of computers and programming in Basic, Fortran and several assembler languages. They also use *DOGS (Drawing Office Graphics System)*, which is concerned with engineering design. General teaching in the department is supported by ten terminals and a fast printer, while DOGS has six terminals and a graph plotter. The Department of Construction uses ten terminals and Management has eight. The Mathematics and Computing Department is in the same building as the Computer Centre which can in theory, support 80 'in-house' terminals.

The structure (see Figure 7.2) is as follows: Linkage to the 'out-departments', which are up to 100 metres from the computer, is at 9600 baud and is based on two methods. One uses a cable consisting of twisted, twisted-wire pairs, which can handle six terminals, with *line drivers* at each end, while the other makes use of a *multiplexer* at each end, the connection being able to suport eight terminals. The in-house terminals are linked by a twisted pair. There is also a link to another college in Slough via a modem and a telephone line which bypasses the college PABX.

The two Harris computers act as if they were one machine and they have common disk and tape drives. The total disk capacity is about 1450 Mbytes, although not all of this is available to users. All communication is through a MICOM 600 port selector switch which sits between the 'Harris' and all the terminals and which routes the terminals to Harris input/output 'ports'.

In practice, one of the multiplexer links to Construction only handles four terminals. System usage is quite heavy and at any time between 8 a.m. and 8 p.m. could be supporting over 60 users with a reasonable response time, say 15–30 seconds. You must bear in mind that an eductional environment is rather different from an industrial or professional one. Students generally will not be highly proficient in keyboard use and since processing is free to them, they must be prepared to put up with occasional highish response times – perhaps 10–20 seconds. At peak times, it sometimes goes up to half a minute, particularly when 60+ terminals are in use.

Remember, it is not just a question of the time to switch between terminals lines and I/O ports. Many students will be writing and running quite large Cobol, Basic, Pascal and Fortran programs. Others will be using various mathematical, statistical and engineering packages or running some of the more complex emulation programs that can make the Harris act like a microprocessor (such as the 6502 or the more modern 68000). At the same time, members of staff will be processing timetables and exam results, and developing and word processing teaching/training material, etc.

Students and staff are allocated user areas on disk for which there are varying levels of protection. All programs are loaded from the system disks and

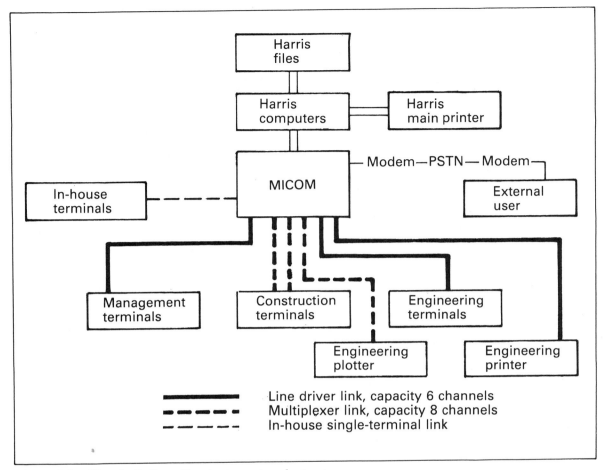

Figure 7.2 *Slough College Harris star network structure*

all data is stored on them so the Harris spends quite a lot of its time reading and writing to disk.

All of this means that a user must expect variable response times. In a 'real-life' situation, anything much over 10 seconds is just not acceptable. Recent studies in the USA have shown that the productivity of professional keyboard users such as experienced programmers or very fast data entry clerks drops off dramatically if the response time goes over 1 second. This may seem ridiculous, but consider a clerk who enters names and addresses in order to build up a mailing list. Companies that do this often have hundreds of thousands on file and they need to have them entered very quickly to avoid excessive costs. The clerks employed press keys as they look at the source document and they develop a rhythmic

approach to the job in which they can almost ignore the terminal. In this way, a very experienced operator can acheive speeds of several thousand keystrokes per hour. If anything happens to disturb the rhythm, such as a hang-up for a while because the system is busy, the operator's flow of movement is seriously disturbed and work rate falls off dramatically.

Tree structure

A tree structure is exactly that: the various terminals are at points on the branches of the tree (see Figure 7.3).

The main difference between a tree and a star is that the star consists of a series of independent

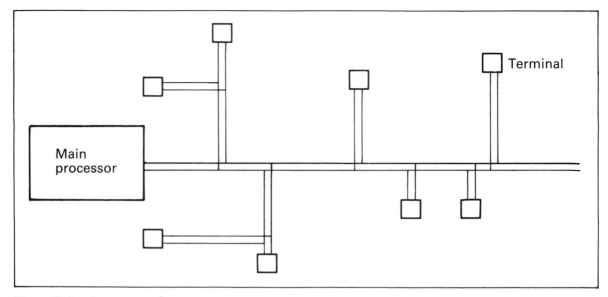

Figure 7.3   *A conceptual tree structure*

links between each terminal and the processor, while with a tree, all terminals share the transmission link. This means that communication is on a *broadcast* basis in which all users listen to messages. Essentially all terminals are competing for use of the medium as well as the files and printers (we say that there is *contention* for the facilities). Each device must be able to recognise messages intended for its own use. However, since the devices (or *nodes* as they are aften called) are independently linked to the processor, breakdown of one does not affect the others. It also means that if the line can be tapped to allow for the join, an extra node can be attached without disturbing the others. A common procedure is for the network to be able to detect that more than one user is 'talking' and to make them stop for a while and then re-start at different times. One basis for this is called *Carrier Sense Multiple Access with Collision Detection (CSMA/CD)*. The line when not actively being used for transmission is in a certain state which is different from when it is carrying a signal or message. If the line is 'busy', other users will not be allowed to receive or transmit, otherwise data from different devices will interfere – collide. The network will also be able to pick up a 'collision' which would occur if two users are 'on' at the same time. (This is often referred to as *Listen before talking to check that the line is free and listen while transmitting to detect a collision.*)

Trees are also called *bus* and *multidrop* structures although these terms really apply to the method of transmission as much the physical arrangement of the LAN components.

Slough College has a number of LANs in use, one of which is a tree based on the *Econet* system (see Figure 7.4) developed by Acorn Computers who were responsible for the BBC microcomputer. The Econet structure allows for a theoretical 255 devices with an effective data transfer rate somewhat reduced below the (again theoretical) maximum, largely because of disk access times and the time to transfer data from disk to the network. The Slough implementation links up ten BBC microcomputers.

Ring structure

This is a very widely used approach where all the components are arranged in a ring. Again, there is contention amongst the users and techniques have been developed to ensure that data reaches the correct receiver and is not damaged by collision with that from another user. Each device will pass messages along to the one next to it unless it recognises one for itself, so each device plays a little part in running the network but the routing structure is simplified – 'if its yours take it and say that you have, otherwise pass it on'.

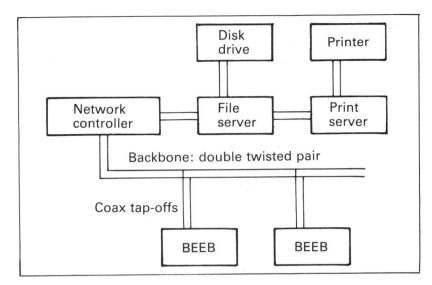

Figure 7.4 *Slough College Econet structure*

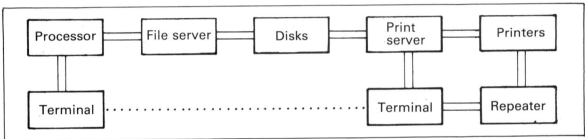

Figure 7.5 *A conceptual ring structure*

Figure 7.5 shows the essentials of a ring containing several terminals. Repeaters may be needed within the ring in order to regenerate signals so that the actual length of the ring can be increased.

A problem with the ring that does not arise with stars and trees is due to the fact that if the ring gets broken at any point, unless steps are taken, the whole system will break down. One way to get round this is to connect alternate devices (see Figure 7.6).

Another problem can be that, since all devices must connect to two adjacent ones, it is rather difficult to arrange for wiring to be laid to accomodate future expansion, unless you know in advance where the devices will be sited.

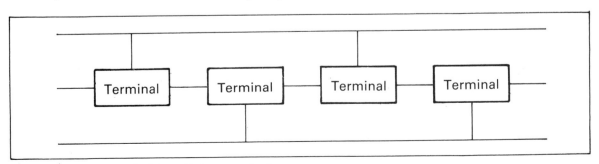

Figure 7.6 *Security linking of alternate devices*

A company called Fibronics offers a 'Cabling Plan' service with which they help users to organise all transmission media including coax and fibre optic cable. It has been shown that over 25% of terminals are moved each year for one reason or another and Fibronics claim to make the best use of existing cabling for expansion and they can provide fibre optic trunking for a whole building.

Variations on the basic topologies are possible, such as a series of loops tapped off a tree (see Figure 7.7).

The essential difference between a loop and a ring is that rings can be designed so that nodes take an active part in network processing where they have sufficient intelligence. In a loop, the controller is the only node that has any control over the use of the channels by the nodes in the loop.

Sometimes the topology of the site will determine the topology and structure of the LAN. For example, it may make sense to arrange that each floor of a building is a loop within a large tree, or perhaps a separate ring for each application, with bridges to others – a manager in a 'spreadsheet/database' ring might bridge across to the 'word-processing' ring to get a report typed.

Once the number of terminals wanted in an LAN goes much above 50, there may be a case for using a PABX, especially since before long the *ISLN* *(Integrated Services Local Area Network)* will appear. With the high-speed switching technology available, it is now economical for digital switches to set up links for quite small packets, such as one line of text within an electronic mail message or a stock balance figure for an order enquiry. LAN suppliers are conscious of the way their markets will go and many products employ or make provision for CCITT X25, IBM SNA and similar protocols.

Another possibility for a PABX is to provide remote access within an LAN or between LANs within a building. For example, an LAN on one floor could act as a 'remote terminal' to an LAN on another floor.

Access techniques

These are not directly of concern to the LAN customer because a particular network will be selected for its general suitability, and although you will not usually have a choice in relation to a particular product, you should be aware of some of the techniques that are employed to enable users to gain access to LAN facilities.

Polling

This is a very common technique in star networks and it means that the central processor goes round a cycle in which each terminal in turn is interrogated to see if it wants to send or receive. In

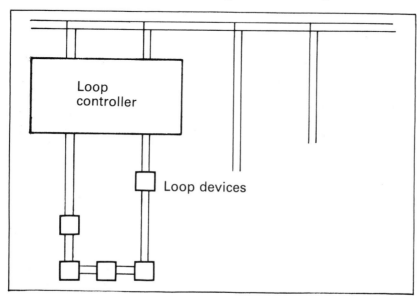

Figure 7.7 *An LAN loop*

practice this is not unlike simple multiplexing. And like multiplexing, performance can be improved if a priority scheme is introduced (something like with a statmux) whereby certain users are given more time slots. The main difference is that multiplexers usually interleave data on a byte (or even bit) basis (Chapter 2), while in the LAN it will be data packets.

Empty slot

Obviously, unless something like FDM is employed, not everyone can 'talk' at the same time. The problem is largely solved in a ring because as far as the processor is concerned, you can talk as much as you like – it will listen when it is your turn. But with a broadcast network or a tree, this would be very inefficient. For trees and some rings, the CSMA/CD technique mentioned earlier is widely used and is satisfactory when data is short and 'bursty', i.e. when there are not too many collisions. The usual problem with CSMA/CD is that the formula used to determine how long a user waits after a collision makes the wait longer and longer until eventually the user device gets fed up and indicates that an error condition has probably occurred. For lowish demand conditions, over 90% of the available channel capacity can be used but this will, of course, drop off dramatically as the user load increases.

The slot technique involves sending one or more empty 'packets' round the ring helped on their way by repeaters. When a device is ready to send it places its data packet into a mini-buffer and as the slot passes, if it is empty, the device inserts its packet into the data area of the empty slot packet, sets an indicator in it to show that it is in use, attaches a destination address and passes it on. The receiving terminal will transfer the data into its own mini-buffer and the packet passes on back to the original sender, perhaps after the receiver has set an acknowledgement indicator in it. The original sender can then be set back to empty once the sender picks it up. One of the problems is that the time to pass the packet round may be so short that its beginning reaches the sender before its back end has been sent. This can be overcome by adding a delaying buffer to the ring and by using a limited number of short packets, but a lowish overall data rate will be the result.

Token passing

This is a development of empty slot in which the signal packet or *token* is seized by a device which attaches data packets behind it. As they come round again, it can remove the data that has been acknowledged and eventually release the token for another user. The Philips SOPHONET broadband system referred to above, uses a token for each channel and, in addition, a particular device can only have exclusive use of a token for a limited time. This makes it easier to predict response times in a busy LAN. The system also has available, a 4-level priority system to make resource allocation more controllable. They claim data transmission rates of up to 30 Mbps.

ALOHA

This is really of historical interest now although it is the basis for CSMA and its derivatives. The 'pure ALOHA' method originated in the University of Hawaii with a large number of terminals spread over the islands tapping into a central computer by radio link. If two data packets collided, the central computer could pick this up by looking at error-check fields in the data. The channel utiliation resulting from ALOHA was very low, but it served its purpose in linking many users cheaply. A development was *slotted ALOHA* in which the central processor allocated time slots and a terminal could only transmit at the beginning of its slot.

Ethernet, Omninet and Novell

There are dozens of LANs and many more suppliers than can possibly be covered in this kind of book but we must mention two products and one supplier in particular since they are related to the majority of small-to-medium LANs in use these days.

Ethernet

This name crops up all over the place in LAN literature. It is a technique developed and now licensed for re-use by Xerox, DEC and Intel. It is probably one of the most successful LANs. Most LANs installed are some derivative of the basic

Field content	Preamble	Destination		Source		Type	DATA	Check field
		Block number	Device number	Block number	Device number			
Byte length	8	6		6		2	46-1500	4

Figure 7.8 *Ethernet packet format*

Ethernet ideas. (The term 'multi-vendor supported system' is often used in the industry to indicate a system which incorporates techniques and ideas from an LAN, like Ethernet products, initially offered by its originator. Econet from Acorn and other LANs are said to have 'single-vendor support'.)

Ethernet is a baseband tree using CSMA/CD, based on coax, and can handle data rates up to 10 Mbps over distances up to 2.5km. Implementations are based on 500-m cable lengths, to which up to 100 terminals can be attached. Up to two repeaters are possible between individual devices and the overall five cable segments support up to 1024 terminals.

Connection of nodes to the main cable is with the aid of a *transceiver*, a buffered device which taps into the cable usually via four twisted pairs and links back to the Ethernet communications controller (which could be a single card in a PC). It could also be a gateway device to other networks or a file or printer server. Ethernet specifies a packet whose length is from 64 to 1518 bytes, the format shown in Figure 7.8.

The preamble contains information that enables the various terminals to get ready to look at packets. The source and destination indicate who sent and who is to receive a packet and also holds the block number which allows devices to be grouped logically: the first bit of the destination block number, if set on, indicates that a block of users is to be spoken to, thus allowing for broadcast messages. The data field indicates the length of the data, which can be from 46 to 1500 bytes, and the final field is for a complex error-checking procedure. Note that there is no specific end-of-frame marker.

A transceiver handles the signalling for transfer and reception of data between devices and the CSMA/CD, while the network controller handles packet formation and deformation.

DEC have recently announced a new range of products including their *Cheapernet*, which is a low bandwidth Ethernet derivative based on thin coax, leading to the same data transfer capacity and much lower cabling costs at the expense of shorter maximum cable length. However, individual LANs can be 'daisy-chained' onto others by processor linkage.

Omninet

This is a very successful single-vendor system offered by Corvus Systems. The medium is twisted wire to a slightly different specification to that usually followed. An appendix to the book deals with protocols and standards and mention will be made of *RS232* and the corresponding CCITT *V24*. These are essentially the same and are followed in most situations where a terminal links to a processor, such as in an LAN. Omninet is based on linkage according to *RS422* or *V11*, which is really an upgrade to RS232 to take advantage of the modern integrated circuits that post-date RS232.

Omninet uses a two-level CSMA and allows for up to 64 devices at 1 Mbps spanning over a kilometre. The circuitry between each device and the network is called a *transporter*. When a request for transmission is picked up, the network waits for a 15-microsecond 'silent' period and informs the host processor that transmission is possible and the relevant transporter is informed. Once a message has been received it is acknowledged and if the positive acknowledgement is not received within a specified time, a retransmission is requested.

During this signalling time, a further transmission may have started and a high-speed carrier sensing is brought into operation. If the carrier is still free, the requested transmission continues. If the line is busy, the new request is denied and has to be re-entered. The procedure is said to eliminate collisions: the terminal actually wanting to send/receive does not have to detect carrier-in-use because all the other terminals will have picked it up and be held back until it is complete.

Omninet was originally implemented with Apple microcomputers, but has been modified for other systems including DEC. Systems are usually configured to make use of one of a range of Corvus disk drives which can hold up to 80 Mbytes, printing and file handling being in the hands of servers.

Novell

This US company seems to be turning into a major force in the PC-based LAN market. Its main products, distributed by Persona (UK) based in Kingston, are a series of LAN software that have their basis in Omninet but are faster and much more sophisticated. Systems use CSMA/CD with positive acknowledgement and the error detection/correction mentioned above. The *G* family of Novell network software *Netware/G* comes in three versions and acts as an operating system sitting over that of the PC. The most powerful version, aided by an Intel 80826 processor, can handle 15 Mbytes of memory and billions of bytes of disk space. All versions support multiple file and print servers, electronic mail, file and record locking and many other features.

The LAN supported is called *G/NET* and can handle up to 255 PCs, each of which is attached to the network by a plug-in adaptor board called a *Local Network Interface Module (LNIM)*, each of which has a Z80B processor with 64K of 'dual-ported' memory. This means that the host computer can write to this memory without in-

terfering with the internal memory of the PC. Also, with clever software, it is possible to arrange for messages to be stacked up in the LINM. This reduces the time delays that would arise if each message or request had to be acknowledged and dealt with as soon as it arrived.

A total basic distance of 4000 feet is possible for the network and optional bridge cards can be attached which extend the length by 4000 feet for each card added.

Upgraded versions of the system, called *G/X25* and *G/SNA* provide PC access to X25 public data networks (such as PSS in the UK) and IBM mainframe-based networks.

Selecting an LAN

The only way to select an LAN is to have a very clear idea of your present and future requirements, exact details of what your current hardware and software is capable of and how much you want to spend. If at the same time, you have picked up enough from this book to give you a broad idea of what you can get from an LAN, you can now talk intelligently to the dealer.

Bear in mind that your problem is to acheive an acceptable level of processing to obtain your results. Considerations such as whether baseband or broadband, data only or voice and data combined, conformity with OSI, etc., are part of the means to the end, not objectives in their own right. So, be advised by your expert. Hopefully, you will have a good idea of what your application wants from an LAN and it will be up to the dealer to put in the best hardware/software for your needs.

Perhaps the first thing to bear in mind is that you don't go for maximum speed and capacity but maximum reliability and reasonable speed. You have presumably already decided that an LAN approach to your data processing is more economical than dedicated PCs or terminal access to a mainframe, so the first thing is to list the applications you want to support and then try to estimate the data volumes and file space associated with each one.

You may find it convenient to make up a table or

Software application	User type	No. of users	Bytes transferred per hour	Storage bytes needed	Hourly characters printed
Electronic mail	Casual	15	500	10,000	500
Order entry	Experienced	5	70,000	500,000	
Word processing	Experienced	3	50,000	300,000	20,000
Word processing	Casual	5	1,500	10,000	2,000
Spreadsheet	Casual	2	10,000	350,000	-

Figure 7.9 *LAN potential user/usage matrix*

matrix as shown in Figure 7.9 for this (the sophisticated manager might care to use his spreadsheet). Although the table seems to show a large number of users, the figures are best worked out for each application. Here, everybody uses electronic mail and some word processing, order clerks may handle customer letters, managers may use the spreadsheet and graphics but might take the odd special order, etc.

Estimates will be totally dependant on the application: for example, the order entry figures in the table were decided as follows. Each clerk is involved with say 25 orders per hour. The keystrokes required to support the order involve first entering the order number and customer number so that the system can carry out a credit check. This process already requires several hundred bytes to flow round the network when you count up the dialogue characters: 'please enter customer number', 'order number not valid for this customer', 'credit available – £300.00' etc. Once credit has been verified, details are entered – catalogue number/quantity/discount, etc. This might take an average of 300–500 keystrokes and a further 1000 bytes from the system. The system will need to read customer, stock and ledger disk records, update them and put them back. At the same time individual clerk files may be updated for security purposes or commission calculations. This leads to an overall figure of perhaps 70,000 bytes per hour per order clerk.

The next problem is estimating file sizes, which really needs professional advice. Determining the size of ledgers, stock files, etc., is not too difficult, but system work space and security back-up files are much more difficult.

The amount of printing is also not as easy as first appears since in this kind of application there will be a need for draft- and letter-quality printing. A typical matrix printer can handle about one page per minute of nice clear printing, except that it looks like printing (you can make out the dots) rather than the more personalised typing that many people still expect. To produce this kind of printing needs a daisywheel or similar printer and these are somewhat slower than the dot-matrix. You will need to decide on the number and mix of printers you need and how long on average the different users can wait for the output to be printed off (spooled) after it has been produced on disk. One possibility is to consider the use of a laser printer such as from Xerox or Canon. These are rather more expensive but very much faster and one of these might take the place of quite a few dot-matrix printers in an environment where users do not need instant printing.

Perhaps the next most important step is to look at the required siting of terminals. Again, the application will often have much to do with this. Presumably the word-processing clerks will be close together as will the order clerks, but other users are not so easy. Is everybody to have a terminal or could several occasional users share one? Will a terminal provide enough extra information or reduce work loads to justify its existance and the extra load on the network?

Having decided on the number of terminals and their general siting, the next item is to think immediately about the future. There is going to be a one-off cost for installation such as cable laying (drilling through walls, etc.), disk drives, printers and the basic network software. It is important to allow for expansion. If you think ten terminals and two printers will do the job, allow for fifteen terminals and three printers even if it means buying a larger capacity disk and splitting off a loop from the basic network. Go for the fastest medium there is – the cost is marginal in the long run. Also bear in mind that a tree uses less cable than a ring and is usually easier to extend.

The next consideration is whether or not you are making use of PCs already and, if so, will any proposed network be able to incorporate them. It is not a good idea to be tied into particular terminal hardware and if you have a few 'weird' micros, it may be worth considering buying a whole set of new ones. This might have been a horror story a couple of years ago when an IBM PC, for example, cost over £2000. Now, there are several very powerful micros on the market including a few IBM PC look-alikes that will be available before this book is published and which offer IBM performance for less than £1000.

Next think about the need for bridges and gateways. Are you just interested in 'desk-to-desk' communication within the office or do you want to go outside? Do you want all users to have access to telex or can needs be met by a dedicated operator? Will viewdata from your (or someone else's) mainframe cut costs and/or generate sales? Are there any existing LANs within the organisation that can be extended for your application, or could there be a requirement to link it into the one you propose to install?

Two final considerations before referring the whole problem to an experienced network designer are to think about the skill level of the users and make a note to investigate the user-friendlines of the software and the amount of training that will be needed. The other is to decide on who to nominate and have trained as the system operator/manager. There are bound to be problems both at the start and throughout the life of the LAN. Bitter experience with mainframes and minis has shown that somebody needs to be dedicated to the system and know enough about it to be able to route problems to the right place even if not actually solve them. There is no difference in principle when the problem is translated into LANs. Once a complex network is put in, users will start getting familiar and before long, try to do 'clever' things which will cause further problems. Someone needs to sort this out as well as monitor the general performance of the network, deal with complaints and suggestions and look to the future.

Appendix One

Mercury

Having spent some time dealing with BT facilities, it is really rather important to say a little more about Mercury. It is not the purpose of this book to detail all the services provided by BT and Mercury, but as a relatively new public telephone company, we must give an overview of Mercury.

By the end of 1982 it had been clearly established that Mercury Communications Ltd were able to offer a complete network that linked fully with that of BT and at the same time could link to international services.

Since then, Mercury has established a series of trunk circuits based on fibre-optic and microwave links, has implemented its *Americall* London–New York link and a London–Hong Kong link via satellite and up to the end of 1986 will have invested about £200 million in its digital-technology networks.

Americall is said to combine the advantages of international direct dialling and leased lines. The charges are based on the time-of-day difference in the UK and the USA.

Northern Telecom are the main suppliers of the larger switches used, while Plessey supply the small-to-medium.

The 'National Operations Centre' in Birmingham is the focal point for a series of optical-fibre trunks, the cables being laid along British Rail inter-city lines. The northern loop brings in Leeds and Manchester, while a southern loop includes London and Bristol. A further extension will cover the South Coast, Wales and Scotland. It is claimed

that well over 500 miles of optical fibre have been laid so far.

Satellite links are available to the US and Hong Kong and extra linkages will be available to several of the areas in which Cable & Wireless has its interests, in particular the Caribbean. (It is no

Figure A1.1 *Mercury distribution node*

Figure A1.2 *Close-up of a microwave antenna at Charing Cross Hospital*

accident that C&W own Mercury since they have enormous interests in undersea cable, satellites, communications services and hardware and complete networks in over 60 countries – in Chapter 3, we mentioned a recent order for Thorn-Ericsson equipment for Hong Kong).

The initial services are based on leased lines; the distribution networks from the trunk network to a city-centre node or building are via optical fibre or microwave and from individual buildings via 'line-of-site' microwave dishes or BT leased lines. Their first microwave city distribution centre was installed in 1983 at the BP building in Moorgate. The single pole carries a microwave horn for 64-kbps services, while the complex antenna is a series of microwave dishes for fast point-to-point services.

The service options are 2 Mbps or 64 kbps, both provided in several combinations.

International communications satellite centres are based in Oxfordshire and in London (at East End Wharf), with Northern Telecom's DMS-250 switch providing the basis.

Network services provided make extensive use of 32-channel lines (CCITT recommendation I421!).

A very advanced network has been set up in the City of London, based on over 150 miles of underground pipework in the City and on the other side of the Thames. This means that access to all of Mercury's high-quality, high-speed services can be obtained very easily and extended cheaply to cover integrated services facilities as they become available.

Basic City Network service options for voice and data include:

- *8 Mbps*: either as a massive data stream or as a multiplex of four smaller channels.

- *2 Mbps*: again, as a single stream or with a number of combinations based on 30, 54 and 60 channels.

- *64 kbps*: in a variety of combinations.

Network maintenance is based on five area maintenance centres and Mercury claim that they support a 24-hour service, with a 4-hour call-out from the time the customer calls in (or the Mercury network monitors pick up the fault).

Trunk circuits are designed so that they are laid in loops with signals routed both ways round it. This means that if a fault is detected, a receiver can pick up the signal from the other end, eliminating the need for back-up circuits. A high error rate in one direction can result in switching and re-synchronising within a quarter of a second.

The Mercury Network Operations Centre (MNOC) in Birmingham operates a highly sophisticated computer-controlled network supervision system. It is called *Supervisory Control and Data Acquisition (SCADA)* and is heavily dependant on the spare channel in the I421 mix.

The main 30 channels are for voice/data. One of the spare channels is for common-channel signalling. The other is for SCADA signals which as well as controlling data/information within the network, can also be used to interrogate and report faults on customer equipment and take action as agreed by the customer.

Data security

Mercury, as one of the most important carriers of data/information, must be sure that it is secure

Figure A1.3 *Installing optical-fibre cable in the City of London*

from 'tapping'. They rely on the facts, that it is almost impossible to break into optical fibres by a physical tap and that there is no side radiation from the cable which might otherwise be picked up. Even more important is the complex way that messages are combined and multiplexed. This means that even if a good tap could be made into a line, it would be next to impossible to decode what was picked up anyway.

A final safeguard, of course, is for the customer to make use of the many data encryption methods that can be applied.

Mercury's *Easylink* involvement with electronic mail and telex has aleady been covered.

Stop press

Three pieces of news from Mercury came in recently as this book was in its final stages, all worth mentioning.

Cheaper calls From August this year, Mercury will be offering cheaper calls for customers directly connected. They claim local discounts of up to 25%, up to 20% savings on trunk calls and up to 17% on certain international routes. (This contrasts strangely with BT's even more recently announced price increases!)

ICL PSS link-up C&W (who you will recall own Mercury) and STC/International Computers have announced that they will shortly set up a joint venture to be operated by Mercury. The new organisation will operate and manage a packet data network in the UK. The service will be called *Mercury 5000* and will initially be based on ICL's existing X25 packet network, to be expanded later on. The claim is that the new network will be the first to offer an *any-to-any* capability, providing and managing connection between differing hardware on a nationwide basis.

Trial satellite data services Later this year Mercury will be evaluating a completely new satellite data service in conjunction with IBM, The London Stock Exchange and Electronic Data Systems Ltd. The service is intended for users who are looking for sophisticated data communication between a relatively few central sites and a large number of out-stations on a nation-wide basis, terminals being equipped with a 1.2-metre dish antenna that can receive from and send to the host computer. Host data is converted from the host hardware protocol to the network protocol and sent at 256 kbps, in packet form. Terminal data is returned similarly at a lower speed. They claim that the service costs will not be too 'distance-sensitive' and are considering expansion into Western Europe.

Appendix Two

Northern Telecom

We have already mentioned this company (in Chapter 3) as the supplier of DMS-250 switches to Mercury, and as the largest supplier of fully digital telecommunications systems in the world, so we must say a little more about them.

The company started as Northern Electric in Canada and in 1976 changed its name and image with its commitment to the *digital world*. They realised that digital technology, although desperately expensive to develop, was the way to ensure the full integration of voice, data and services allowing for future developments and trends in the volatile and very sensitive computer and telecommunications markets. So, although their competitors felt that the technology was possibly too difficult and expensive to develop, Northern set out to become (their words) 'the first telecommunications manufacturer in the world to offer a complete line of fully digital switching and transmission equipment'. They seem to have acheived this objective, particularly after going heavily into the US marketplace and taking on the giant AT&T, with a gross world revenue (1985) of about 5,700 million Canadian dollars. Published figures claim systems involving over 27 million fully digital lines throughout the world. Currently, they employ about 47,000 staff in nearly 50 countries and supply large and small switches, optical fibre (over 400,000 km by the end of 1985), communications systems and office products. They expect to plough back about 2 billion dollars into R&D by 1989. (In common with many large companies, Northern find that a new product range can easily involve development costs of £200 million.)

A separate subsidiary company, Bell-Northern Research (70% Northern) was formed with Bell Canada to carry out R&D and it currently has 4,700 employees in 10 laboratories, integrated on a world-wide basis, which share resources and information in product development. The main responsibility of Northern's recently opened site in Maidenhead is the development and support of DMS switches and this kind of support and know-how was probably the conclusive factor in obtaining the Mercury business.

Maidenhead also plays a part in an international DMS program for Cable & Wireless in the Caribeean, a major telephone system in Turkey and Mercury in the UK. The central site for this R&D is in Ottawa where the company information database is stored under what they call *PLS (program library system)*. High-speed data links connect this to Maidenhead and other UK centres as well as three US laboratories and all sites share and update the databases. Another aspect of Maidenhead is what they call three 'captive offices', systems that can simulate any installation, from a small office switch to the largest DMS-250 set-up. They are used in integrated development and training programs and for tailoring switches to customer requirements. This is particularly important since it helps Northern to maintain their 'evergreen' policy for products, whereby developments and extensions can be delivered to existing customers as 'batch change supplements', making switches upwards-compatible. This means that customers can take advantage of major developments without existing equipment becoming obsolete. In 1976, they introduced their *SL-1* switch of which they have since sold or have on order, over 1900 throughout the world. The *DMS-10* was launched in 1977 as a small digital central office switch for systems with up to 5000 lines and in 1979 they launched the *DMS-100* family of large central office switches that handle up to 30,000 lines (the 1000th of these was recently installed as the hub of a communications centre for Los Angeles International airport). Another product in this range is the *DMS-300*, an international gateway switching centre. A derivative of the DMS-100, named *SL-100*, has been used successfully in the USA for military, university and business applications and is now being sold in the UK. (GEC produce it under licence with the name *ISLX*). BABT approval is expected by the time this book has beeen published.

These later switches are intended for use in 'integrated services networks' and Northern base their network technology on what they call the *OPEN* (*O*pen *P*rotocol *E*nhanced *N*etwork) *WORLD*, a philosophy and technology back-up that shares out to maximum effect the use of digital transmission and switching capabilities that can exist within a network providing the necessary 'connectivity' between different users in different applications with different hardware.

Centrex is a business service option (now available in the UK) based on DMS250, half for the business applications and half for trunked lines. The DMS100 is used for local exchanges and SL100 provides a large PABX for local use within the system. Recently, they have demonstrated their commitment to ISDN by setting up major field trials in Bermuda where a DMS100 supports 200 lines at 64 kbps and in Arizona, Oregon and Toronto, where two 300-line systems are undergoing tests at up to 160 Kbps.

As well as providing most of the switches for Mercury, Northern also supplies British Telecom and has recently negotiated a contract with Barclays Bank for an SL-10-based digital private network which will be worth £40 million over the next five years. The bank will have a national communications system based on the OPEN network concept. Support for Barclays is at Maidenhead, in the hands of Northern's Communications Systems Division.

Other recent contracts have been for the world's longest public optical-fibre systems in Canada using cable and opto-electronics supplied by Northern's transmission group and Canada's largest provincial telecommunications network based on $11 million of switching equipment including two SL-200 and nine SL-1 switches.

Finally, Northern supply a rather different product called *LRS-100* which is fundamental to a £20 million British Telecom program for customer service improvement. The system has already been installed at sites throughout the UK and supports BT's Repair Service Centres. Before very long, 10 systems should be handling about five million exchange connections. The system provides an automatic testing and fault analysis capability which is handled by a mainframe computer whose database has basic data on the characteristics of each terminal equipment type, i.e. telephone. The system can send signals to a subscriber unit which tests various electrical properties such as resistance, capacitance, etc., and compares them with the recorded standards, both to diagnose reported faults and to anticipate faults that might arise.

Statistics collected are added to the database and in its most sophisticated form can be used for service billing purposes and even to organise service calls with the customer and local repair centre.

Northern manufactures and assembles the system at Hemel Hempstead and they are pleased to claim that they make the maximum involvement of local industry such as a £400,000 contract to Passim for the supply of disk drives and multiplexers for LRS-100 systems. To date Northern Telecom has installed more than 80 LRS-100 systems worldwide. These automate customer services for more than 14 million exchange connections operated by 30 telephone companies in 6 countries.

Appendix Three

Value-added networks

As we said at the beginning of Chapter 3, there are only five licensed PTOs – British Telecom, Hull Corporation, Mercury, Cellnet and Vodafone. The significance of this is quite simply that at present no other company can offer a network service that *only* provides for the carrying of telecommunications messages. Any potential competitors will need to be licensed in the same way and be subject to the same regulations and control as the present five PTOs.

However, there is nothing to prevent a company trying to offer some kind of service that makes use of existing networks as the 'medium' for the service. Hence the term *value-added service* or *VAS*, which covers the facilities provided by companies, based on the network of a PTO, over and above the actual carriage of messages. However, the term *value-added network (VAN)* is much more commonly used.

(You might care to note that the term 'value-added' is now becoming an established commercial buzzword. For example, in a recent radio programme, a manufacturer was referring to the increasing trend toward sales of value-added meat products (presumably pre-cooked, pre-prepared, pies, etc.) compared with raw meat itself.)

VAN facilities offered

These involve the following which may be combined:

Data This includes all the various services that are geared to central computers providing a database and access to it. It could also include individual computer processing, i.e. remote use of applications software, technical and commercial. In particular, it includes the viewdata services detailed in Chapter 6, such as Prestel itself and the Prestel Citiservice (which is aimed at investors and commodity traders). A different application is in credit card validation and electronic funds transfer. BT offer *CRAFTS* which can check out the balance supporting a credit card transaction and then transfer credit from cardholder to credit card company. Similarly, home-shopping should be included here, such as the Birmingham 'Club 403' which allows for supermarket purchasing. Many other businesses are available through Prestel.

Since the bulk of these services are involved with the processing and recording of business transactions, they are actually referred to as 'transaction services'.

The term 'managed' data network is rather different and means the provision of a service on which other companies can base their own data network. In particular, it provides routing/re-routing and error handling at a higher level than is available from the basic network.

Text This involves the electronic mail detailed in Chapter 5, but in particular such services as BT Gold and Mercury's Easylink and not forgetting telex itself which is not generally considered as a VAN facility because it is actually supplied by BT and Mercury as part of their existing networks. There is no reason in principle why other companies could not offer a similar value-added service.

Often the electronic mail service is nothing much more than a 'store-and-forward' or 'deferred transmission' facility, but it could include 'broadcast' options where one message is relayed to many receivers.

Although not actually electronic mail, some VANs provide word processing which, of course, allows users to format the electronic mail that they want to send.

Another possibility is the provision of a fax service (such as that from BT) that will send fax images for a company that does not have a machine and convert fax images to a more acceptable form if required.

One point that needs to be brought up to remind you of Chapters 5 and 6 is the possible confusion over viewdata and electronic mail. Strictly speaking, as you can see from Chapter 6, viewdata is page-oriented in a format that usually resembles that of Prestel.

We could confuse the issue further by calling services that offer access to large text libraries as 'electronic publishing', particularly Prestel and Dialog which are mainly concerned with convenient access and retrieval of information pages.

Insurance   BT is involved with two services for the insurance industry. *Mediat* is in use by about 100 insurance brokers with several different terminals supported. The system operates over the PSS via Multistream and allows quotations for motor and life, pension and mortgage plans. Many other facilities can be provided and BT claim that over 95% of business users are within local-fee access of the network.

A similar service being tested by 16 companies is called *PINS (Prestel Insurance Network Services)*.

Voice   Here we are refering to services involving not just the transmission of voice calls, but their redistribution after recording, in other words, voice 'libraries'. This is something like a large-scale answering machine service. (It is interesting to speculate on whether people will refer to this kind of service as 'audiotex', being somewhat similar to videotex.)

BT offer *TAN* which is an audio message answer-ing service mainly used by direct response advertisers and *VOICEBANK* which is a 24-hour message board facility, a sort of audio electronic mail. An exciting extension has already appeared involving data entry and enquiry via a telephone dial or keypad from which audio tones, i.e. sound rather than voice, are sent to the system, and a range of recorded voice response. Voicebank can be accessed similarly. This means that a booking system, for example, could be geared to two kinds of user. The first would be where clients ring up and leave their order on the 'voice bank' to be scanned regularly by the dealer. The second is more sophisticated and means that the client must learn the code that enables the receiving computer to know what is wanted. The client can be guided by means of voice responses from a pre-recorded library or even from a speech synthesiser.

Yet another area that falls into this category is advanced teleconferencing facilities, although with ISDN this will be available as part of the 'regular' network.

Image   This is a good point at which to make an apology. The subject of telecommunications is incredibly wide and expanding fast. It was felt that some areas had to be omitted and video is one that was chosen.

The least we can say is to mention that video-conferencing is something that quite a few companies feel is worthwhile. Similarly, 'slow-scan' TV. The problem with sending TV pictures is that for very clear definition such as might be wanted for photographic purposes, high-quality, high-speed lines are needed. We can expect this to be possible as ISDN hardware develops (there has already been some experimental work carried out with fibre-optics giving transmission rates of between a half and one billion bps). For the present, BT and Mercury can support a lower rate TV transmission.

Advantages of VANs

In reality, a VAN does not necessarily give that much more than a user can provide for himself, except that it is cheaper and rather more wide-ranging. This is exactly the same as applies to computing facilities in general. Where you have a specific computing requirement that is customised to your own peculiar requirements, you are unlikely to be able to adapt a general-purpose facility. But when you want the same in principal as everyone else, somebody will supply the service.

Take a very simple voice-response library. Imagine you have a small number of stock lines in which a client can order or ask about stock levels by part number. As well as setting up the transmission network, you will need programs to handle the processing and the recognition of what is wanted. You will need to set up a voice library containing the descriptions of all parts sold, as well as words like 'Sorry', 'Out of stock', 'Not stocked, try again', etc. A company supplying this kind of WAN service will do most of the set-up for you, saving both time and money.

Because of the economies of scale involved, the typical user will have access to many functions that would just not be available otherwise.

Sharing resources with many other users will reduce overall costs and should guarantee the integrity of the service. These days, with VANs being so much in mind, the suppliers are under a lot of pressure to provide an efficient and reliable service. On the other hand, it is not difficult to become 'locked in' to a particular VAN and this could possibly become one of the most important considerations for the future: the level of standardisation. Having set up and become used to a particular system, if you are dissatisfied you may have a major re-orientation problem in getting used to a new system. With the increasing adoption of the OSI model in network design, a user will eventually be able to change terminal equipment or even VAN supplier with, hopefully, no great problems.

The general level of communication both within a company and with its customers/suppliers can be improved. Voice response, for example, could be used for customer ordering, but it could also be used to provide instructions and directions for a mobile sales force and one of the major benefits is less paper being circulated.

Regulation and trends

The Department of Trade and Industry have granted over 180 licences to date, involving well over 600 different licenced services, electronic mail and its derivatives accounting for about one third.

The rules relating to licences are, as with telephone services, designed to protect the consumer as well as the PTOs themselves, but have given rise to some confusion and have led to problems. In particular, it has been found difficult to classify a particular service offered or even if it classes as a VAN at all.

Registration procedures are not well defined and nor are the fee-paying mechanisms.

The DTI has issued a working document that suggests the creation of a *Managed Data Network Service (MDNS)* licence to companies that don't really want to go through the rigmarole of applying for a full VAN licence. This may add to the confusion because it could be said that the error-handling and management statistics provided by MDNS does add considerable value to the existing transmission network. The only area that does seem to be confusion-proof at present is the way that the 'basic conveyance' from BT and Mercury has been defined. This will provide a good framework for the future, since eventually, the DTI will have to consider applications for PTO licences from other companies.

Appendix Four

Protocols and standards

Apart from the convergence of voice and data and the leaps towards ISDN, probably the most recurring theme in this book has been that of standards (or lack of them in a particular area) and it seems appropriate at least to try and summarise the main areas, especially those specified by CCITT and ISO.

Where possible, relevant standards have been introduced in the appropriate chapters, e.g. fax groups 2–4, and will not be mentioned here.

Perhaps the most important set of ISO standards, those relating to OSI (Open Systems Interconnection) are the subject of Appendix 5.

You should also be aware of the American IEEE (Institute of Electronic and Electrical Engineers), the other major body that sets standards. The whole IEEE range cannot be covered here but examples include, specifications 802.3 and 802.2 which relate to OSI layer 1 and data link layers, respectively, while 802.4 relates to an LAN token bus scheme.

The CCITT 'V' series

There are about 50 of these recommendations, which relate to data transmission over telephone circuits. Some are highly specific and relate to particular applications, but those in the V21-29 range are very important. Note by the way that several are apparently duplicated with the suffix *bis*. This means a modification or extension to the one before it. Similarly, in the case of V27, the suffix *ter* means that a second extension has been applied.

The following is a list of the more common, lower-numbered recommendations:

V21 300-baud modem specification standardised for switched telephone network use.

V22 1200 bps full-duplex, two-wire modem specification standardised for use in a general switched network.

V22 bis As for V22 at 2400 bps.

V23 600/1200 bps modem specification standardised for use in a general switched network.

V24 Probably the most well-documented standard in the whole business. It corresponds loosely to the EIA (US Electrical Industries Association) specification *RS232* (current revision level C) and is one of the very few standards generally accepted by hardware manufacturers. It defines in detail a suitable interface between terminals (DTE – data terminal equipment) and modems (or DCE – data circuit-terminating equipment). The specification includes the mechanical characteristics of the plug and socket and which signals go with which pins, the electrical characteristics of the signals, a description or the signals themselves and details of the standard subsets of the recommendation itself. For example, the various lines of circuits have specific functions: circuit 103, for example, is used to send data from DTE to DCE, while 104 sends data in the reverse direction. Other details include how the DCE informs the DTE that it wants to send or is ready to send. (Specification V24 corresponds to layer 1 of the OSI 7-layer model discussed in the next appendix.)

V25 This relates to automatic call/answer equipment in a general switched telephone network.

V26 2400 bps modem for use on four-wire point-to-point circuits.

V26 bis 2400/1200 bps modems standardised for a general switched telephone network.

V27 4800 bps modem for leased circuits.

V27 bis 4800/2400 bps modem with automatic adaptive equaliser standardised for leased circuits.

V27 ter 4800/2400 bps modem standardised for a general switched network.

V29 9600 bps modem for use on leased circuits.

The CCITT 'X' series

There are quite a few of this series of recommendations, of which the following are just a selection:

X1 International user classes of service in public data networks.

X2 International user facilities in public data networks.

X3 PAD facility in a public network.

X20 Interface between DTE and DCE for stop/start transmission services on public networks.

X21 General-purpose interface between DTE and DCE designed for interface with synchronous, V-series modems.

X24 List of definitions of interchange circuits between DTE and DCE on public data networks.

X25 Interface between DTE and DCE for packet terminals on public data networks.

X28 DTE/DCE interface for a stop/start terminal accessing a PAD on a public network.

X29 Procedures for exchange of control information and user data between a packet terminal and a PAD.

X50 Fundamental parameters of a multiplexing scheme for the international interface between synchronous networks.

X80 Interworking of inter-exchange signalling systems for circuit switched data services.

X96 Call progress signals in public data networks.

A further list of X series recommendations appears at the end of Appendix 5.

Miscellaneous protocols

Various other 'X' recommendations have been mentioned *in situ* and several more appear at the end of Appendix 5.

A *Data Link Control* is a set of rules to ensure that data is transmitted when the communication channel is available and the sender is ready to receive

An early standard, developed by IBM from a basic 'stop/start' protocol was *BSC* or *Bi-Synch (Binary Synchronous Communication)*. This has been largely succeeded by *SDLC (Synchronous Data Link Control)* also by IBM. It is essentially a hardware-link specification that relates to two-way transmission of data on a single channel.

SDLC is in many respects compatible with *HDLC (High-level Data Link Control)*, a set of ISO standard linkage protocols which are used heavily in CCITT X25.

A very important set of standards is IBM's *SNA (Systems Network Architecture)*. This is a general set of protocols for terminal–terminal and terminal–computer-network control. It was initially developed for host-terminal linkage, but was later extended to handle terminal–host–host–terminal linkage and eventually the fully meshing structures needed to support the fully distributed processing that is now so important. SNA, although quite different from OSI (Appendix 5), shares many of its features, in particular a layered structure. It has acheived a wide acceptability because of the deep entrenchment of IBM in mainframes, front-end processors and microcomputers.

Numerous VANs (Appendix 3) operate with the linkage of separate SNA networks.

Only time will tell whether SNA or the ISO open system connection will become the more widely used: it seems likely that usage of the two will develop in parallel until an ever more universal approval can be adopted.

Appendix Five

The ISO 7-layer model

The International Standards Organisation has proposed an *Open Systems Connection (OSI)* standard, a framework around which rules and a common approach can be set up for the intercommunication and interconnection of computers and terminals of all degrees of dumbness. This is not actually a protocol specification, but a set of ground rules or a guide that should be followed if a hardware or software designer hopes to link units (lines, terminals or even networks) with different characteristics. Any potential conversation can be considered at different levels and the ISO model identifies seven, each of whose activities and services require varying degrees of protocol. It assumes that 'peers' (processes or processors at the same level) will follow the conventions recommended in order to communicate successfully, irrespective of the supplier of hardware or software.

The levels are called *layers* (see Figure A5.1) and involve the dialogue between entities at the same level, such as the user and the applications software, two networks, or switching or control stations within a network.

Layer no.	Usage	Application
7	Application	To enable the user to link to lower levels
6	Presentation	Conversion between different codes; formatting data
5	Session	The 'conversation' rules between users
4	Transport	Ensuring that the network service is upheld
3	Network	Interconnecting users
2	Data link	Link between user and network
1	Physical	Actual form of bits transmitted

Figure A5.1 *The OSI 7-layer model*

The actual characteristics of the physical medium that makes up the network are not specified because they will vary so much. For a wide area, it includes the coax, microwave or optical-fibre link, any intermediate modems, packet assembler/disassemblers, multiplexers and so on. LANs can be so complex as to be almost indistinguishable from a WAN and the increasingly widespread use of networking and inter-networking means that private ISDN will develop more quickly than public. Since connection can be between networks, LAN and WAN and between different terminals and computers within and outside a particular network, suppliers will need the same standards template as for WANs. Even though the ISO model was specified some time before 'wide area LANs' were recognised as being so important, it can be applied to any networking facility, whether communication is destination-directed or broadcast.

The levels in this ISO model are as follows.

1 Physical layer

The actual connection and control of bit-string messages from the sending/receiving device DTE (Data terminating equipment which could be a terminal or the computer itself) to the network (physical medium), in particular the modem that hooks the digital data into the analogue network. The protocols will relate to the physical/electrical properties of the terminal and the link such as whether baseband or broadband, the modulation technique and the voltage/current corresponding to 1/0, etc. They may also include the procedures for the initial contact and eventual sign-off.

2 Data link layer

This relates to the reliance that must be placed on the actual link as to its ability to cope with (detect and possibly recover from) noise, faulty transmission/reception and amongst other aspects, includes parity checking, checksum, etc. In addition, it involves the formation of data 'frames', i.e. the arrangement of data bits and the control signals needed to pass them on. It also relates to collision detection, packet acknowledgement and retransmission and the control of data flow rates.

3 Network layer

Set up to cover the switching, addressing and routing of individual message packets between the various 'nodes' and terminals within the network and inter-network connection. The packet address defined by layer 3 is used in the next layer to ensure that packets in a different network are correctly received.

4 Transport Layer

This involves the efficiency of data transfer between systems that differ fundamentally, such as networks provided by different suppliers, and requires protocols that allow the different networks to use their individual characteristics and qualities to the best advantage in order to maintain 'quality of service'. It also applies to path selection when alternatives are available. In particular, it is here that checks are carried out to make sure that data sent has been received properly and in the right sequence. Data will be received via higher levels, packetised, sent to the other end and reassembled in the correct order. The top three layers are different from those below because they do not relate to the network, but rather to what it is being used for. In other words, intercommunication rather than interconnection or, in yet another way, they relate to the user not the network supplier. (The term 'end-to-end' is often used to refer to these layers.)

5 Session layer

A session is the total time and the various functions carried out when one facility is connected to another. At the beginning of the session, the two ends will 'negotiate' over details like:

- Who speaks first?

- Will continuous, two-way transfer occur?

- Will conversation be alternately one-way in bursts, etc?

- When does the session end?

The negotiation is part of the establishment of connection and itself will be the subject of protocols. It could also relate to certain activities which, while not exactly part of the application, nevertheless, make the application easier to manage. For example, when transferring text as documents or other data as whole files, various markers might be agreed on so that one end can detect end-of-page or end-of-file within the bulk of the data. General dialogue formats will be the concern of the next level but layer 5 will be concerned with controlling the interplay of messages and data within the defined dialogue format.

When considering a network or a mainframe computer as a complex data source, rules will be · needed as to which service requests take priority. Similarly, queuing and buffering arrangements may need negotiation.

6 Presentation layer

The two parties may 'speak different languages' in that data representation codes may be different. This could be a simple matter, where perhaps the receiver requires data in EBCDIC while the sender is generating ASCII.

At a higher level it could even be a rationalisation of local differences between real languages. The service being sought by a user might be the translation/execution of programming language programs, or database enquiry using an enquiry language. The negotiation will need to establish the way these differences are overcome.

Similarly, rules relating to page or document format must be laid down for videotex or teletex.

7 Application layer

This is really the personal layer in that it provides an interface for the application programs (word processing, fax, viewdata or whatever) and the actual data collection/reception point, such as a human end-user or a graphics terminal. It is here

that we could also consider certain network management and support functions such as accounting for facilities so that users can be charged or, at least, costs be related to various cost centres.

Another possibility might be the gathering of network performance statistics such as traffic levels and intensities, queuing and waiting times, breakdown and hang-up logging and so on. In fact, the upper layers are becoming blurred together by technology. Several companies offer very sophisticated hardware whose principal function is to make data transfer efficient and cost-effective. In particular, a concentrator could be used to bulk up data for retransmission at a much higher speed and this is technically tied up with the transport and network layers. Several companies now offer highly intelligent concentrators which contain ROM-based accounting and statistics-gathering software. These free the terminal processor to concentrate on the raw data processing that forms the application.

There are currently not very many standards/protocols relating to the levels above 3.

'X' series OSI-related recommendations

A number of the 'X'-series are aimed directly at OSI, including:

X200   The reference model specification itself

X210   OSI layer service definition conventions

X213   Network service definition

X214   Transport service definition

X215   Session service definition

X224   Transport protocol specification

X225   Session protocol specification

Appendix Six

Glossary of useful terms

Accoustic coupler A low speed modem (typically 300 Baud) that uses a telephone handset to transmit and receive data by means of sound and that does not need to be permanently connected to the communications line.

AM (Amplitude Modulation) A technique for transmitting analogue data in which a 1 or 0 affect the height of a carrier signal differently.

ANSI (American National Standards Institute) This body specifies U.S. standards such as ASCII.

ARQ (Automatic repeat request) A transmission procedure in which the receiver automatically sends a request for re-transmission if a message is improperly received.

ASCII (American Standard Code for Information Exchange) A digital coded set which represents each character of the standard typewriter keyboard as a 7-bit digital code.

Asynchronous transmission *See* Stop/start.

Attenuation A decrease in signal strength during transmission due to loss of power through equipment, lines, or other transmission devices.

Autodial/Autoanswer A modem that does not require a telephone to link a terminal to the telephone line.

BABT (British Approvals Board for Telecommunications) The government body responsible for ensuring that telecommunications

equipment is of a standard that allows it to be connected to networks offered by the PTOs.

Bandwidth The range of frequencies that a communications line or channel can carry; the wider the bandwidth, the higher the baud rate.

Baseband A transmission technique in which a digital signal is modulated with a carrier, the carrier then being used just to transmit the one message.

Baud One signalling element per second. A measure of the signalling rate of a line, i.e the rate at which it can switch between electrical states. The data transfer rate, measured in bits per second (bps), is usually a simple multiple of this.

Bit A 1 or 0 usually represented electrically, the (Binary Digit) elementary unit from which data and computer instructions are built up.

BCC (Block Check Character) Used in cyclic redundancy, a BCC is a character accumulated and transmitted by the sender after each message block. This is analogous to the block check character that is accumulated by the receiver to determine if the transmission was successful.

BPS (Bits Per Second) A measure of data channel information transfer rate.

Bridge A device that links two widely separated local area networks so that they appear to be one network spanning a large geographical area.

Broadband A transmission technique in which

several digital signals are modulated with a single carrier signal. In this way, several messages can be accomodated on one channel.

Broadcast A method of transmission in which all stations are capable of receiving a message transmitted by another station.

BSC (Bi-Synch) Short for binary synchronous communications. An IBM procedure that uses a standardised set of control characters and control character sequences for synchronous transmission of binary-coded data between transmitter and receiver. The synchronisation of characters is controlled by timing signals generated at the receiver.

Buffer An intermediate storage area within a computer or a network that is used to link devices of differing data send/receive rates.

Bus A high-speed high-capacity cable used within a computer.

Byte A group of eight bits representing one data character.

Carrier sense *See* CSMA/CD.

CCITT (Consultative Committee for International Telephone and Telegraph) An international standards group for communications standards and recommendations.

Ceefax The teletex system broadcast by the BBC.

CEPT Conference Europeenne des Administration des Postes et des Telecommunications.

Channel A means of transporting information signals. Several channels can share the same physical circuit.

Character code Any organised code for representing graphics and control characters. e.g. ASCII, EBCDIC.

Circuit switching A communications switching system that completes a fixed circuit from sender to receiver at the time of transmission.

Close coupling An LAN structuring technique in

which dumb terminals utilise the processing power of an intelligent 'host' for network facilities.

Cluster Two or more terminals or computers connected to a data channel at a single point, managed as if a single system.

Coaxial cable A cable consisting of one conductor, usually a small copper tube or wire, within and insulated from another conductor of larger diameter, usually copper tubing or copper braid.

Collision Detection *See* CSMA/CD.

Common channel signalling A method where the signalling for a number of channels is sent over a dedicated channel, separately from the messages themselves, typically 1 signalling channel for 30 voice/data channels.

Compression A process that reduces the number of bits required to specify some data, usually a pixel array. Also, a means of converting a set of repeated characters into a single character and a repetition factor to reduce the amount of data to be transmitted.

Concentrator A network node that stores data for later re-transmission at a higher rate.

Contention Unregulated bidding for a computer facility or communications line by multiple users.

Control Character A character, other than a graphic character, whose occurrence in a particular context initiates, modifies or stops a control operation. e.g CR/LF (carriage return/line feed) to denote 'end-of-line'.

CSMA/CD (Carrier Sense Multiple Access with Collision Detection). A broadcast transmission procedure in which the system can sense when more than one user is trying to send a message. Once such a 'collision' has been detected, appropriate action is taken to try and ensure that it is not repeated.

CRC (Cyclic Redundancy Check) A technique used for data transmission error control in which extra data is added which is mathematically related to the content of data to be sent.

CUG A facility in Prestel, in which a number of

users can be treated as if they were an internal group, communication being possible only between members of the group.

Database An organised collection of data on a computer storage device. The application does not determine the way in which the information in a database is structured so users can view it in a manner pertinent to their individual requirements for processing or information retrieval.

DCE (Data Circuit-terminating Equipment). CCITT term for a class of data devices under the control of the supplier of the network.

Decompression Expansion of compressed data into its original form.

Dedicated line A communications channel leased from a PTO for the exclusive use of a private concern.

Dial-up service A remote computer facility which requires a telephone and modem connection, initiated by manual dialling.

Digitisation The conversion of analogue data such as sound or a telephone signal in a digital form by sampling or averaging over a period of time.

Download The process of sending data or program instructions from one device in the network to another device. e.g. downloading a computer game from Micronet for later use. Can also apply to files, where for example, a user might download a set of figures from a viewdata service for later spreadsheet processing.

DTE (Data Terminal Equipment) CCITT term for the data terminal devices themselves, a category which includes the computer and VDU.

DTMF A set of pairs of audio tones that represent the digits 0–9 used in modern telephone connection. CCITT and Bell Telephone have set the two main standards.

Dumb terminal A terminal that has no facility for data processing other than for the handling of input and output.

Duplex (Full duplex) A circuit or a protocol that

permits transmission of a signal in two directions simultaneously.

EBCDIC (Extended Binary Coded Decimal Interchange Code). The main IBM character code set for mainframe and minicomputers.

Echo check A method of checking the accuracy of transmission of data in which the received data is returned to the sending end for comparison with the original data. (Sometimes called read-after-write.)

ECMA (European Computer Manufacturers Association) This group plays a vital role in establishing European standards.

EFT (Electronic Funds Transfer) The transfer of credit for business use by data transmission.

EIA (Electronic Industries Association) This group has established many electronic interface standards including RS232C.

Electronic mail A non-interactive communication between users that is transported electronically, not physically.

Ethernet The Xerox-originated baseband LAN technology.

Fax (Facsimile transmissions) A system of communications in which a document, photograph, or other hard copy graphic material is scanned, and the information transmitted to a remote receiver, where the image is reconstructed and duplicated onto paper.

Fax groups The CCITT recommendations for different levels of facsimile transmission. Group 1 relates to very slow, low-resolution fax, Group 2 is faster, Group 3 is faster still with better resolution and Group 4, which is still under development, is very fast and intended for use in packet networks.

FM (Frequency Modulation) A technique for transmitting analogue data in which a 1 or 0 affects the frequency of a carrier signal differently.

Frame A set of encoded data containing all the necessary data to reconstruct the sampled data for all channels of a digital transmision system.

Gateway Services that provide compatibility between systems of differing network or communication architectures. It converts functions from one into a form recognised by another, for example, accessing telex through Gold.

Glass Teletype The name given to a dumb terminal that includes a VDU.

Half-duplex A circuit or a protocol that permits transmission of a signal in two directions, but not both directions at the same time.

Handshake A communication set-up process in which sender and receiver exchange signals to acknowledge each other's existance and establish compatibility.

Hayes commands Aset of standard commands related to autodial/autoanswer modems originally developed by the US Hayes company, now being adopted by may other modem suppliers.

HDLC (High-level Data Link Control) An ISO Data Link Layer bit-synchronous communication protocol for data links. Similar to SDLC.

Hertz Abbreviated to Hz. The measure of frequency, being equal to one cycle per second.

Host Any network node that a user can access for processing power, information/data files, and applications software.

Host computer A mainframe or minicomputer on which applications can be executed and which also provides a service to local and remote users of a computer network.

IEEE (Institute of Electrical and Electronic Engineers) The US professional organisation that also defines standards, such as the IEEE 802.3 standard for LANs.

Intelligent terminal A terminal that has some processing power over and above that required for dealing with input and output.

Intelsat The INternational TELecommunications SATellite.

Interface A boundary between two devices or two separate pieces of software for which the form and functions of the signals that pass across it are standardised.

Internetworking Operation involving the interworking of two or more networks.

IP (Information provider) An organisation that provides viewdata, etc., and which has a responsibility for creating the data, formatting it for the user and keeping it up to date.

IPSS The international packet-switched service.

ISDN (Integrated Switched Digital Network) A conceptual fully digital network which allows access for all users to a wide range of facilities.

ISO (International Standards Organisation) An organisation that sets standards in many areas relating to telecommuications and data communications.

ISO layered model *See* OSI layered model.

LAN (Local area network) A communication system intended to support exchanges of data and commands between stations within a confined area, such as a single building, a factory, a university site, etc.

Leased line A communications line providing a permanent connection between two nodes. The customer leases or rents the line from a PTO.

Line driver A device used in a line which increases signal strength by regenerating the signal rather than amplifying it, since this would increase the noise too.

Line protocol The detailed procedure for the exchange of signals between the source and the receiver, designed to accomplish message transmission.

Loose coupling A LAN structuring technique in which intelligent terminals share network resources with an intelligent host, but which can carry out their own processing.

MAN (Metropolitan Area Network) A network that spans a complete urban area.

Microwave transmission A technique for transmitting data through the air using a carrier signal of very high frequency sent/received by means of a dish aerial or a horn.

Modem Device that converts digital information into an analogue signal suitable for sending over analogue phone lines and then reconverts the analogue signal once received into digital information.

Modulation A number of different techniques used to enable the transmission of digital data in an analogue form.

Multiplexer A device that divides a data channel into two or more independent fixed data channels of lower speed.

NAPLPS (North American Presentation Level Protocol Syntax) ANSI standard for Videotex/Teletext transmission of text and graphics.

Network topology The general arrangement of terminals, nodes and hosts in a network. Examples include a ring, tree and star.

Node A point at which two or more communications circuits meet. Commonly used in computer networks to describe a point where processing is performed such as a concentrator, amplifier or cluster controller.

Oracle The teletex system broadcast by independent TV.

OSI Layered Model The ISO Open Systems Interconnection Reference Model. A conceptual scheme in which the functions common to network architectures are arrayed in a layered hierarchy. The model has also been adopted by the CCITT.

Packet A collection of data to be transmitted, typically containing routing and error-correction information.

Packet terminal A terminal which has the necessary hardware and software to generate/receive packeted data, thereby being able to communicate directly with a PSS without going through the PAD.

Packet switching A process in a data transmission network that is designed to carry the data in the form of packets. Each packet is passed to the network and the devices within it use the header information to transmit the packet to the correct destination. Packet Switching System is abbreviated to PSS.

PAD (Packet Assembly/disassembly Device). Permits terminals that do not have an interface suitable for direct connection to a packet switched network to access such a network. As well as converting the terminal's usual data flow to and from packets, the PAD handles all aspects of call set-up and addressing.

Parity checking A technique of error detection in which one bit is added to each data character so that the number of 'one' bits per character should always be even (or always odd).

PCM A time division modulation technique of great importance in digital telephony, in which the analogue signal is sampled and the values obtained are quantised as a binary value which can be represented by an audio tone.

Polling A method of controlling terminals on a multipoint or clustered data network where each terminal is interrogated in turn by the computer to determine whether it is ready to receive or transmit. Data transmission is only initiated by the computer.

Prestel The name given to the British public viewdata service. Also the name of the graphics protocol used by that service and adopted by private viewdata suppliers.

Protocol A set of rules agreed to by two communicating parties for accomplishing some specific set of tasks.

PSS *See* Packet switching.

PSTN (Public Switched Telephone Network) The familiar (voice) telephone system over which calls may be dialled.

PTO One of the (currently) five national, government-licensed suppliers of telecommunications services.

Quantisation During PCM (*q.v.*), once a sample value has been obtained from an analogue signal, it must be converted into a digital value. This is done by comparing it with one of a limited number of values in the possible range and rounding it off to the nearest. This quantisation can produce some distortion because when the original analogue signal is regenerated, the quantisation error due to the rounding off cannot be recovered.

Repeater A device used to extend the length and topology of the physical channel beyond that imposed by a single segment, up to the maximum allowable end-to-end channel length.

Ring structure An LAN topology in which all nodes can only connect with the two nodes adjacent, each node routing for the others.

Routing The ability of a network node to pass on messages from a different node to another different one.

RS232 The IEEE standard for the interface between computers and slow devices. Corresponds very closely to CCITT V24.

Sampling A technique used in digitisation in which data values are taken at selected times or averages are taken over selected time periods.

SDLC (Sychronous Datalink Control) An IBM communications line discipline or protocol associated with SNA.

Signal-to-noise ratio The ratio of the level of the desired signal to the level of the noise associated with it at a given time. The ratio is usually expressed in decibels.

Simplex Transmission that is in one direction only, for example a television or a data entry terminal.

SNA (System Network Architecture) IBM's network architecture for mainframe/terminal communication. Functionally similar to the OSI layered model.

SPC (Stored Program Control) The name given to an exchange or PABX ·which carries out its functions under the control of a computer-like processor.

Statistical multiplexer Device that divides a data channel into a number of independent data channels greater than that which would be indicated by the sum of the data rates of those channels. It does this on the basis that not all the channels will wish to transmit simultaneously, capacity thus being freed to accomodate additional channels. Often given the name *Statmux*.

Stop/start A serial data transmission method in which each character is transmitted as a selfcontained set of state changes, i.e. as units needing no additional timing information.

Synchronous transmission A transmission technique in which electronic state changes are carried out under the control of a 'clock' (timing mechanism).

TDM (Time division multiplexing) A method of combining several messages onto a single channel in which equal time intervals are allowed for each.

Teletex A passive, one-way service where a series of sequential screens appears on a monitor without interaction with the user. It is more like closed-circuit television with "stills" rather than "action". Oracle and Ceefax in UK.

Teletext A European standard for text communication and electronic mail, introduced as an enhancement and eventually intended as a probable replacement for telex.

Telex The public switched low-speed (telegraph) data network which is used worldwide for the transmission of messages.

Terminal A device for sending and/or receiving data on a communication channel. Typically a VDU/keyboard, but could be input only, such as a timecard reader or output only, such as a printer.

Token-passing An LAN ring technique in which a special piece of data is circulated so that terminals can request services from the network.

Tree structure A network topology in which each terminal is connected to the host by tapping off a single main channel.

Trunk A communications channel between two switching devices.

V24 Probably the most well-known CCITT 'V' recommendation, it defines in minute detail an interface between DTE and DCE.

VAN (Value-added network) A network offering more than just the facility to provide for the interconnection of messages.

Videotex *See* Viewdata.

Viewdata A two-way information access/retrieval system that uses a video monitor or TV as a display device.

Virtual circuit A logical path through various elements in a communication system, established for purposes of conducting some specific exchange between two or more parties. (Of particular importance in networks based on the X25 protocol such as PSS).

Voice-grade line A term applied to channels suitable for transmission of speech and digital or analogue data or facsimile, generally with a frequency range of 300 to 3000 Hz.

VT57/VT100 A range of display terminals from DEC whose associated protocols have become standards de facto.

WAN (Wide area network) A communications network for use over long distances.

Window On a VDU screen, a portion of the display set aside for a specific purpose or application and cut away from or overlaid on the rest of the display.

Word processor A program run on a microcomputer (sometimes designed just for the purpose) that makes the entry, correction and amendment of text easy – in effect a highly sophisticated typewriter.

Work station An intelligent, input/output terminal in a LAN or WAN usually associated with data processing; more than just data entry and output.

X25 The international standard protocol from CCITT for accessing packet-switching networks. The networks support virtual circuits between network access points.

X200 The series of ISO standards relating to OSI.

X400 The series of CCITT standards that relate to electronic message handling systems (mail).

Index